Competing in the Face of Climate Risks

AI Disclosure Statement: Microsoft Copilot, Open AI's ChatGPT, Anthropic's Claude, and the World Bank's mAI platform were used selectively during the preparation of this report between October 2024 and May 2026 to support editing, summarization, and rapid synthesis of the literature and illustrative examples. In Chapter 1, AI tools were also used to support work related to the evaluation of corporate communications data, including analysis of the diffusion of green technologies and changes in environmental regulatory pressures across markets. All substantive text was written by the authors. All references, numerical information, and factual claims were individually checked and verified by the authors, and all AI-assisted output was reviewed and revised through multiple rounds of human editing.

Reproducible Research Repository
https://reproducibility.worldbank.org

A reproducibility package is available for this book in the Reproducible Research Repository at https://reproducibility.worldbank.org/index.php/catalog/study/FR_WLD_2026_668.

Scan to see all titles in this series.

MIDDLE EAST, NORTH AFRICA, AFGHANISTAN, AND PAKISTAN DEVELOPMENT REPORT

Competing in the Face of Climate Risks

Evidence from Firms and Policy Priorities in MENAAP

Zeina Alsharkas, Jean Michel N. Marchat, Enrique Aldaz-Carroll, and Ana Goicoechea

WORLD BANK GROUP

1 2 3 4 29 28 27 26

ISBN (paper): 978-1-4648-2335-0
ISBN (electronic): 978-1-4648-2336-7
DOI: 10.1596/978-1-4648-2335-0

Artificial Intelligence: If AI tools were used in the development of this work, page ii of this work will have further information.

Cover photo: © Nancy Pauwels / Shutterstock. Used with the permission of Nancy Pauwels / Shutterstock. Further permission required for reuse.

Cover design: Bill Pragluski, Critical Stages, LLC

The Library of Congress Control Number has been requested.

The manufacturer's authorized representative in the European Union for product safety is Mare Nostrum Group B.V., Doelen 72, 4831 GR Breda, The Netherlands. Email: gpsr@mare-nostrum.co.uk.

Middle East, North Africa, Afghanistan, and Pakistan Development Report Series

This series features major development reports from the Middle East, North Africa, Afghanistan, and Pakistan region of the World Bank, based on new research and thoroughly peer-reviewed analysis. Each report aims to enrich the debate on the main development challenges and opportunities the region faces as it strives to meet the evolving needs of its people.

TITLES IN THE MIDDLE EAST, NORTH AFRICA, AFGHANISTAN, AND PAKISTAN DEVELOPMENT REPORT SERIES

Competing in the Face of Climate Risks: Evidence from Firms and Policy Priorities in MENAAP (2026) by Zeina Alsharkas, Jean Michel N. Marchat, Enrique Aldaz-Carroll, and Ana Goicoechea

Ten Thousand Steps in Her Shoes: The Role of Public Transport in Women's Economic Empowerment (2024) by Muneeza Mehmood Alam and Lisa Bagnoli

Informality and Inclusive Growth in the Middle East and North Africa (2023) by Gladys Lopez-Acevedo, Marco Ranzani, Nistha Sinha, and Adam Elsheikhi

Exports to Improve Labor Markets in the Middle East and North Africa (2023) by Gladys Lopez-Acevedo and Raymond Robertson

Blue Skies, Blue Seas: Air Pollution, Marine Plastics, and Coastal Erosion in the Middle East and North Africa (2022) by Martin Philipp Heger, Lukas Vashold, Anabella Palacios, Mala Alahmadi, Marjory-Anne Bromhead, and Marcelo Acerbi

Distributional Impacts of COVID-19 in the Middle East and North Africa Region (2021) edited by Johannes G. Hoogeveen and Gladys Lopez-Acevedo

The Reconstruction of Iraq after 2003: Learning from Its Successes and Failures (2019) by Hideki Matsunaga

Beyond Scarcity: Water Security in the Middle East and North Africa (2018) by World Bank

Eruptions of Popular Anger: The Economics of the Arab Spring and Its Aftermath (2018) by Elena Ianchovichina

Privilege-Resistant Policies in the Middle East and North Africa: Measurement and Operational Implications (2018) by Syed Akhtar Mahmood and Meriem Ait Ali Slimane

Jobs or Privileges: Unleashing the Employment Potential of the Middle East and North Africa (2015) by Marc Schiffbauer, Abdoulaye Sy, Sahar Hussain, Hania Sahnoun, and Philip Keefer

The Road Traveled: Dubai's Journey towards Improving Private Education: A World Bank Review (2014) by Simon Thacker and Ernesto Cuadra

Inclusion and Resilience: The Way Forward for Social Safety Nets in the Middle East and North Africa (2013) by Joana Silva, Victoria Levin, and Matteo Morgandi

Opening Doors: Gender Equality and Development in the Middle East and North Africa (2013) by World Bank

From Political to Economic Awakening in the Arab World: The Path of Economic Integration (2013) by Jean-Pierre Chauffour

Adaptation to a Changing Climate in the Arab Countries: A Case for Adaptation Governance and Leadership in Building Climate Resilience (2012) by Dorte Verner

Renewable Energy Desalination: An Emerging Solution to Close the Water Gap in the Middle East and North Africa (2012) by World Bank

Poor Places, Thriving People: How the Middle East and North Africa Can Rise Above Spatial Disparities (2011) by World Bank

Financial Access and Stability: A Road Map for the Middle East and North Africa (2011) by Roberto R. Rocha, Zsofia Arvai, and Subika Farazi

From Privilege to Competition: Unlocking Private-Led Growth in the Middle East and North Africa (2009) by World Bank

The Road Not Traveled: Education Reform in the Middle East and North Africa (2008) by World Bank

Making the Most of Scarcity: Accountability for Better Water Management Results in the Middle East and North Africa (2007) by World Bank

Gender and Development in the Middle East and North Africa: Women in the Public Sphere (2004) by World Bank

Unlocking the Employment Potential in the Middle East and North Africa: Toward a New Social Contract (2004) by World Bank

Better Governance for Development in the Middle East and North Africa: Enhancing Inclusiveness and Accountability (2003) by World Bank

Trade, Investment, and Development in the Middle East and North Africa: Engaging with the World (2003) by World Bank

All books in the Middle East, North Africa, Afghanistan, and Pakistan Development Report series are available for free at https://openknowledge.worldbank.org/handle/10986/2168.

Contents

Boxes

Figures

Maps

Tables

Acknowledgments

This report was prepared by a multidisciplinary team from the Prosperity Vice Presidency and the World Bank's Middle East, North Africa, Afghanistan, and Pakistan region under the direction and guidance of Sandeep Mahajan, Nadir Mohammed, Cheick-Oumar Sylla, Denis Medvedev, Asya Akhlaque, Jasmin Chakeri, Sébastien C. Dessus, Mario Guadamillas, Djibrilla Issa, Eric Le Borgne, and Yira Mascaró.

The report was led by Zeina Alsharkas, Jean Michel N. Marchat, Enrique Aldaz-Carroll, and Ana Goicoechea, with key contributions from Susanne Åakerfeldt, Josue Banga, Alexandra Campmas, Lucio Castro, Euijin Jung, Maryla Maliszewska, Mariem Malouche, Anouk Pechevy, Marc Tobias Schiffbauer, and Nolan Young. The team also wishes to thank Reda Aboutajdine, Amira El-Bidawi, Arlan Brucal, Massimiliano Calì, Weronika Celniak, Maksym Chepeliev (Purdue University), Nicolò Dalvit, Fatma Elashmawy, Carolyn Fischer, Matias Julian Garibotti, John Gabriel Goddard, Gal Komem, Camilo Mondragón-Vélez, Eduardo Jimenez Sandoval, Abhishek Saurav, Andrius Skarnulis, Alan Fuchs Tarlovsky, and Alastair Thomas for their valuable inputs.

Feedback from Syed Adeel Abbas, Meskerem Brhane, Kevin Carey, Jan von der Goltz, Laurent Gonnet, Stéphane Hallegatte, Jocelyn Jabbour, Matteo Morgandi, Richard Record, Sonya Sampson, Ashok Sarkar, and Ernest John Sergenti is gratefully acknowledged.

About the Authors

Enrique Aldaz-Carroll is a senior country economist for Namibia and previously served as senior economist in the Prosperity Vice Presidency's Global Trade team at the World Bank, focusing on trade and climate issues. Before that, he was senior economist to the managing director of development policy and partnerships, coauthoring the World Bank Group's Green, Resilient, and Inclusive Development approach and the Strategic Framework for Knowledge. Enrique has also held senior economist positions in Cambodia, Cyprus, and Poland, working on macroeconomic issues. He led the team of 25 World Bank staff that supported the reunification negotiations in Cyprus under United Nations auspices. He joined the World Bank in 2004 as a Young Professional. He holds a master's degree in economics from the London School of Economics and a doctorate in economics from the Institute of Development Studies at the University of Sussex.

Zeina Alsharkas is a senior country economist for the Syrian Arab Republic in the Fiscal Policy and Growth Global Practice of the World Bank's Middle East, North Africa, Afghanistan, and Pakistan region. She joined the World Bank in 2020 as a Young Professional and has since worked as a country economist for multiple countries, including Iraq, Jordan, and Syria. Her areas of expertise include competition policy, firm innovation, productivity, and growth. Before joining the World Bank, she worked as a lecturer in economics at the University of Essex. She holds a doctorate in economics and a master's degree in management economics from the University of Essex, as well as a bachelor's degree in economics from Damascus University.

Ana Goicoechea is a senior economist in the Small and Medium Enterprise and Enterprise Development Unit of the World Bank's Trade, Competition, and Business Department. Her research focuses on the constraints and policy solutions that drive firm growth and productivity, with particular attention to firm capabilities, technology adoption, market linkages, access to finance, and regulatory efficiency. She leads the Firms and Climate Policy initiative, where she examines how climate change affects firm performance and develops policy responses to strengthen business resilience and competitiveness. Previously, she led the Competitiveness Policy Evaluation Lab, overseeing a portfolio of approximately 40 pilots and impact evaluations on private sector development. She brings more than 20 years of experience across the World Bank Group—including with the Infrastructure Network, the Sustainable Development Network, the International Finance Corporation's Investment Climate Department, and the Trade and Competitiveness Global Department—and has worked across a wide range of country contexts. She holds a master's degree in public policy and international development from the University of Maryland.

Jean Michel N. Marchat is a lead economist in the Trade, Competition and Business unit of the World Bank's Prosperity Vice Presidency. Currently, he is leading research work on investment climate, green policies, and agricultural value chains; assists World Bank teams in the design of lending operations; and provides policy advice to countries worldwide. Previously he worked in the Middle East and North Africa and Sub-Saharan Africa regions on trade facilitation, and export and small and medium enterprise development while leading major analytical work such as Regional Competitiveness Reports with the World Economic Forum. Before joining the World Bank Group, in the early 2000s, he worked for several years in Kazakhstan as an adviser for macroeconomics and forecasting in the Ministry of Economy under a European Union assistance program. He holds a doctorate in economics and a master's degree in development economics from the Centre for Studies and Research on International Development, France, and has also attended executive education courses at the Harvard Kennedy School.

Executive Summary

BACKGROUND

The Middle East, North Africa, Afghanistan, and Pakistan (MENAAP) region is highly exposed to climate risks. Countries across the region are already among the most water-scarce in the world and increasingly vulnerable to heat extremes, droughts, floods, and sea level rise. These interlinked hazards are undermining agriculture,threatening food and water security, and imposing mounting economic and social costs across the region. Rising temperatures are already reducing labor productivity, straining health systems, intensifying desertification, and contributing to water scarcity, crop losses, and higher energy demand for cooling across the region, while water scarcity, droughts, flooding, and sea level rise are deepening fragility, poverty, and displacement risks in the region.

This high and rising exposure creates an urgent need to respond. Climate risks are already materializing and affecting economic activity, firm performance, trade, and livelihoods. Delaying action would further magnify losses.

Responding to climate risk can generate benefits that go well beyond reducing related losses. Policies that strengthen access to finance, improve the business environment, upgrade infrastructure, and promote resource efficiency can raise productivity, enhance competitiveness, support job creation, improve health outcomes, and increase resilience to other shocks. In this sense, climate-responsive reforms can deliver a double dividend by addressing both climate vulnerabilities and long-standing structural constraints in MENAAP economies.

IMPACTS OF RISING HEAT ON FIRMS AND TRADE

Against this backdrop, this report focuses on two concrete and immediate examples of how the evolution of climate and related policies affect firms and trade in the region and how timely responses can generate economic opportunities if key policy changes are implemented: first, higher temperatures affect firm performance through reduced labor productivity and increased costs. Second, new climate mitigation policies in the European Union (EU), a key trading partner, will also raise costs for firms exporting to that market.

These two developments present both challenges in terms of lowering the competitiveness of firms that do not adapt and opportunities for economic expansion for those that do. Given that temperatures will likely rise further and that European climate mitigation policies will shortly be fully implemented, swift policy action is needed to adapt.

Impacts on Firm Performance

The analysis of firm-level data from seven countries in the region over 2019–24 reveals that the impact of extreme heat on businesses' performance is significant. Nearly 32 percent of firms in these countries have been facing over 100 days of extreme heat annually—days averaging above 35 degrees Celsius—resulting in lower productivity and higher operational costs. More frequent periods of extreme heat are associated with a 6 percent decrease in sales, a 4 percent reduction in labor productivity, and an 8 percent decline in wages. Smaller firms and those outside global value chains were more affected.

The negative impacts of extreme heat are more pronounced for firms operating in challenging business environments and those with limited access to finance. Firms facing the highest regulatory burden saw sales decline by 17 percent as extreme heat days increased—more than three times the drop experienced by firms with a moderate burden. Similarly, firms with the most severe financing constraints faced a 21 percent drop in sales, compared with just 7 percent for those with moderate constraints.

Despite the negative impact of rising temperatures, the data reveal that few firms are taking measures to adapt. In the Arab Republic of Egypt, for example, only 9 percent of firms invested in cooling technologies, indicating a lack of awareness, incentives, or resources for climate adaptation. Limited access to finance helps explain underinvestment in cooling, with each unit of improvement in a composite access to finance index increasing the likelihood of investment in cooling by 5 percent. Despite energy subsidies, high up-front costs and complex regulatory processes continue to deter firms from implementing adaptation measures.

Adaptation in the region is further complicated by widespread energy inefficiencies. MENAAP countries consume significantly more energy per unit of output than their southern European peers. The analysis shows that, with rising heat, firms increase their energy demand and face higher water supply insufficiency—inefficiencies that can further increase their operating costs. These findings suggest that efforts to improve the business environment, expand access to finance, and invest in infrastructure could be more impactful on firm performance when targeted to firms facing extreme heat shocks.

Impacts on Export Markets

Regarding external markets, MENAAP exporters will need to adapt not only to the rising temperatures but also to changing market access into Europe owing to new climate policies. The EU and UK (United Kingdom) are implementing climate policies to promote a green transition. Key among them are the EU Carbon Border Adjustment Mechanism (CBAM), which entered fully into force in January 2026, and the UK CBAM beginning in January 2027.

The EU and UK CBAMs will impose fees on MENAAP exports of affected sectors based on their emission intensity, which will affect their competitiveness in the European market. The overall impact on MENAAP economies will be limited because the affected exports have a small weight in their economies. Modeling results project that the impacts of the EU CBAM on the gross domestic product (GDP) of MENAAP countries are negative in most countries, albeit negligible, with the MENAAP region projected to experience a net GDP loss of US$1.4 billion, representing only a 0.1 percent decline relative to the baseline. Only Jordan and Morocco are projected to see net GDP gains—of US$20 million (a 0.05 percent increase) and US$21 million (a 0.02 percent increase), respectively.

However, the impact on some sectors could be large because some countries are significantly exposed. For instance, Algeria and Egypt are expected to face high EU CBAM fees in their iron and steel sectors owing to their relatively high carbon intensity compared with the EU and other countries. As a result, their primary iron and steel exports to the EU are projected to fall respectively by 68 percent (US$4 million) and 78 percent (US$14 million), although this represents a relatively small share of their total exports.

Conversely, other countries could potentially gain competitiveness and increase their exports. These include the United Arab Emirates (US$1.2 billion, a 12 percent increase) in primary aluminum, and both Egypt (US$844 million, a 123 percent increase) and Morocco (US$2.2 billion, a 405 percent increase) in fertilizer, assuming they can certify the carbon intensities of their firms.

Impacts on FDI

Foreign direct investment (FDI) inflows into MENAAP have been adjusting in anticipation of CBAM implementation. FDI is an important driver of investment and technological upgrading; however, the CBAM-affected sectors in MENAAP—such as aluminum, cement, fertilizer, and iron and steel—receive relatively little FDI overall. With few exceptions, even countries with lower emission intensities in these sectors rely minimally on FDI. Since the announcement of CBAM, EU FDI into MENAAP's aluminum and steel sectors has declined, whereas Chinese FDI has risen. As a result, the realization of these sectors' potential in the short term will depend primarily on domestic investment.

The EU and UK CBAM regulations pose both risks and opportunities. They will raise compliance costs for MENAAP exporters, possibly reducing profits and market share in Europe, especially for small and medium enterprises with limited capacity to meet requirements. Some firms may need to shift exports to less desirable markets, and FDI could move to more CBAM-compliant regions. Conversely, these regulations could provide market opportunities and attract increased FDI for MENAAP firms with low emissions and firms integrated into EU value chains that comply with environmental and human rights standards.

KEY RESPONSES FOR FIRMS AND GOVERNMENTS

How can firms and governments react to this changing external environment? Improving access to finance (A2F) and the business environment, especially for smaller firms, will be critical. By tailoring A2F initiatives and business environment reforms to firms' vulnerability, governments will help boost productivity while reducing heat-related performance losses. Finally, supporting firms' "green competitiveness" by fostering decarbonization will be critical to ensure future market access.

Supporting Firms

Adapting to increased temperatures and being able to lower carbon emissions will require firms to adopt greener technologies. Technology adoption is a complex process that relies on many factors, such as the expected returns of a new technology, the readiness of the firm (in terms of human resources and managerial capacities), competition and the regulatory landscape, information on available technologies, and the availability of financing.

Financing is critical. A2F for "green" purposes will require developing new financial products (such as green loans, green bonds, sustainability-linked bonds, transition bonds, and insurance products); implementing specific new tools such as green taxonomies; improving domestic savings mobilization; and fostering deeper financial intermediation.

Improvements in the business environment will also be needed to ensure the proper working of market mechanisms to foster adaptation and decarbonization. Interventions in this area will require defining and implementing appropriate regulations such as green standards and conformity assessments; simplifying licensing for heat-resistant buildings or heat adaptation investments, reporting, and disclosure requirements; and imposing environmental taxes and fees.

These actions can eventually be framed as part of industrial policy packages. Supporting eco-industrial parks is another promising policy approach, as these parks can boost resource efficiency, reduce emissions, and foster heat adaptation mechanisms, all while enhancing competitiveness.

Finally, supporting competition and ensuring a level playing field between public and private firms will be critical in MENAAP. State-owned enterprises (which often operate high-emitting sectors) and banks, through their dominant positions in the region, may sometimes slow adaptation and decarbonization by limiting market mechanisms or being slow to react.

Supporting Green Competitiveness

Adherence to emerging climate-related trade regulations is essential for sustaining competitiveness within the European market. MENAAP countries have limited time to prepare and cope with these new trade regulations. Swift government action will be needed in four areas.

Reduce nontariff measures (NTMs) and strengthen quality infrastructures. Reducing tariffs and NTMs on imports of green capital goods, intermediate inputs, and services, as well as strengthening national quality infrastructures relating to environmental and social standards, are crucial. Lowering tariffs and NTMs will encourage the importation of green goods and facilitate the adoption of green technologies.

Advancing quality backbone infrastructure—including metrology, certification, and accreditation—and raising green and social standards will enable firms to prove compliance with European requirements. Strengthening carbon certification will enable firms to trace and certify their carbon content, ensuring that CBAM payments are based on their actual carbon emissions rather than on higher default values. Upgrading quality infrastructure will lower the cost of compliance for individual firms.

As domestic regulators raise environmental and social standards applied to domestic products and processes, meeting the export market requirements will become less costly because the regulatory distance will be smaller. Over time, higher domestic standards will lower the cost of compliance with Corporate Sustainability Due Diligence Directive requirements for firms connected to EU value chains.

Improve incentives to attract green FDI. Sectors can be supported to adopt higher standards and lower-emission technologies through technology-neutral incentives and more targeted ones for technologically mature items (such as high-efficiency air conditioning appliances, high-efficiency water treatment, solar panels, and batteries); transparency measures; and access to green finance. Instead of subsidizing specific technologies, governments can link incentives to measurable energy savings or verified emissions reductions. Mandating transparency on energy use through energy audits and disclosure requirements nudges firms to identify and act on cost-saving opportunities without dictating specific actions.

Governments can also ensure access to low-cost financing for efficiency investments (for example, through risk sharing with commercial banks) so capital constraints do not bias firms toward short-term, inefficient choices. Revenue recycling into technology-neutral support (such as tax credits, accelerated depreciation, or concessional finance for efficiency investments) can also be pursued to avoid picking winners.

Help firms and workers to adjust. The restructuring of affected sectors will require assistance to firms and workers to speed up the adjustment process and reduce negative impacts. This includes assisting sectors in aligning with the environmental and social standards of trading partners and helping workers who lose their jobs to transition to new jobs.

Pursue fossil fuel subsidy reforms and gradually introduce a carbon tax to lower CBAM fees. This would enable governments to retain part of the carbon revenue that would be otherwise transferred to the EU or UK. The resulting funds could be sizeable. For example, in the case of Egypt and the EU CBAM, simulation results show that if the country were to phase out fossil fuel subsidies in addition to implementing a carbon tax, revenues could increase by 4–5 percent of GDP. This could boost public revenues and yield significant societal benefits, including job creation, improved health outcomes, and greater equity, subject to social protection policies being implemented to protect vulnerable firms and populations.

Opportunities for the MENAAP Region

Economic opportunities will arise from the necessary move to more sustainable economies, provided key policy changes are implemented. Adaptation needs will drive demand for sustainable cooling and battery production, which are already attracting foreign investment. New European trade policies may also create trade prospects for countries able to export products with lower carbon intensity than the EU or UK average. Rising renewable energy needs will, in turn, create opportunities for investment in renewables, particularly in solar and wind generation. Indeed, renewables provide a strong competitive advantage for the region. Finally, greater resource efficiency, through reduced grid inefficiencies, energy-efficient buildings, and improved production processes, can boost competitiveness in external markets.

The shocks analyzed in this report will have strong implications for the MENAAP region and its enterprises. They may have some negative impacts on firms and affect select sectors, but they can also offer strong opportunities. These shocks call for immediate and resolute interventions to reduce these negative impacts and enjoy the opportunities for the benefit of the populations in the region.

Abbreviations

A2F	access to finance
AI	artificial intelligence
AVE	ad-valorem equivalent
BOS	businesses of the state
CBAM	Carbon Border Adjustment Mechanism
CCDRs	Country Climate and Development Reports
CGE	computable general equilibrium
CO_2	carbon dioxide
CO_2e	carbon dioxide equivalent
CPAT	Climate Policy Assessment Tool
CS3D	Corporate Sustainability Due Diligence Directive
ECTs	earnings call transcripts
EIP	eco-industrial park
ETS	Emissions Trading System
EU	European Union
EV	electric vehicle
FDI	foreign direct investment
FFLs	fossil fuels
GDP	gross domestic product
GHG	greenhouse gas
GT	green taxonomy
GTAP	Global Trade Analysis Project
GVC	global value chain
IAZ	industrial acceleration zone
IEA	International Energy Agency
IPPU	industrial processes and product use

MENAAP	Middle East, North Africa, Afghanistan, and Pakistan
NACE	Statistical Classification of Economic Activities in the European Community
SMEs	small and medium enterprises
SOB	state-owned bank
SOE	state-owned enterprise
tCO_2e	tons of carbon dioxide equivalent
TEI	Trade Exposure Index
VAT	value added tax
WBES	World Bank Enterprise Surveys

CHAPTER 1

The Impact of Temperature Changes on MENAAP's Firms and Adaptation Policies

INTRODUCTION

Average temperatures and the number of "extreme heat" days—with maximum temperature above 35 degrees Celsius—are increasing rapidly across the Middle East, North Africa, Afghanistan, and Pakistan (MENAAP). Between 1990 and 2023, the region's average temperature rose by 1.35 degrees Celsius, and the number of extreme heat days increased by 11.5 per year.[1] These trends are predicted to decrease output and jobs not only in agriculture but also in manufacturing and services. With the years 2023–25 recorded as the three warmest years to date (WMO 2026), the frequency and intensity of extreme temperature events are expected to rise (Seneviratne et al. 2021), adding further risk to private sector operations in MENAAP. Key channels through which extreme heat may affect economic activity include lower productivity, higher production costs, and disruptions of infrastructure services and supply chains.[2]

Despite the likely adverse effects of rising heat and droughts, many firms in the region still neglect to invest in adaptation measures (Heger et al. 2022; Islam et al. 2022, 2023; World Bank 2022c). Adaptation initiatives seem to receive a lower priority compared to other investment needs.

Market and policy failures largely explain such underinvestment. The uncertainty about the economic costs of rising temperatures (including measurements of gradual impacts on firms' workers, operational costs, and suppliers) is an important explanatory factor. Moreover, many firms may also lack the capacity to adapt because of narrow profit margins, short-term planning horizons, or (perceived) low probability of climate effects and related events (Downey et al. 2023). Business environment policy constraints such as limited access to finance, high costs of doing business, and inadequate infrastructure further raise the costs for firms to invest in adaptation, exacerbating the effects of the market failures (Crick et al. 2018; Grover and Kahn 2024). Understanding the magnitudes and main channels through which rising heat affects the performance of different categories of firms in MENAAP countries, along with the main market and policy failures preventing firms from sufficiently adapting, is key for designing effective policies.

In this context, this chapter attempts to quantify the impact of rising temperatures on firms, assess the adaptation strategies they have pursued, and identify climate adaptation policy options. This quantification is performed by first combining granular climate and firm-level data and then by analyzing earnings call transcripts data. The chapter begins by briefly examining the key recent economic trends shaping the private sector, which provides context for understanding how temperature evolution intersects with broader market forces. It then delves into how heat is directly affecting firms, with a focus on the vulnerabilities stemming from physical hazards and firm-specific exposures. Additionally, the chapter assesses firms' resilience through their adoption of climate-adaptive technologies, evaluating both the extent of these investments and their effectiveness. Finally, it explores the role of policies and regulations in supporting adaptation to heat-related risks, with particular attention to how the regulatory environment and financial access can play a role.

Although both droughts and rising temperatures pose significant risks in MENAAP, this chapter focuses on the effects of extreme heat on firms. The World Bank's Country Climate and Development Reports (CCDRs) across the region consistently highlight rising temperatures as an important threat, with wide-ranging impacts on health, productivity, and economic activity. Extreme heat can intensify drought conditions and affect a broader set of firms, including those in both humid and arid areas. In addition, recent research has already examined how droughts affect firms (Gatti et al. 2024; Gatti, Onder et al. 2025), allowing this report to complement those findings by focusing on heat-specific risks.[3]

A SLUGGISH PRIVATE SECTOR

As of early 2026, MENAAP is facing multiple conflicts having a dramatic human and economic impact, with repercussions around the globe. Regional growth (excluding the Islamic Republic of Iran) is projected to decelerate from 4.0 percent in 2025 to 1.8 percent in 2026 (Gatti, Özden et al. 2026). This reflects a sharp slowdown in the Gulf Cooperation Council (GCC) economies, where growth is expected to decline to 1.3 percent in 2026 (Gatti, Özden et al. 2026). For GCC and Iraq's economies, a key concern is the disruption to their production, processing, and export of oil and natural gas. Beyond this, the region has struggled with persistently low growth since the 2000s, underperforming global peers. A key explanatory factor is a weak private sector, which remains stagnant and unable to drive sustainable economic expansion (Gatti, Onder et al. 2025).

The region's private sector productivity has been declining and remains below global benchmarks. Firm-level data from nine countries in the region over 2013–23 show that sales per worker shrank by 8 percent, which is partially explained by weak investments in capital goods and human resources: only 22 percent of firms invested in physical capital and 15 percent offered formal training (Gatti, Onder et al. 2025).[4] Consequently, MENAAP economies underperform their income peers in product and process innovation, with lower spending on research and development (Gatti, Onder et al. 2025). Low productivity is compounded by weak competition due to the dominance of state-owned enterprises, distortionary regulations, and uneven taxation, all of which discourage firms from increasing productivity and formalizing (Lopez-Acevedo et al. 2023).

Additionally, structural issues, such as widespread informality and gender disparities further hinder overall productivity. Informal firms (accounting for 40–80 percent of total employment, depending on the country) are typically less productive than formal firms yet remain key competitors (World Bank 2022c). Furthermore, women's workforce participation is just 18 percent, and only 3–14 percent of firms have female top managers, compared with 21–22 percent in middle-income economies overall, leading to missed economic opportunities in the region and significant economic losses (Gatti, Onder et al. 2025).

Significant business environment obstacles indeed constrain growth, limit competitiveness, and shape key features of the private sector. Access to finance is a prominent challenge cited by most firms in Afghanistan, the Arab Republic of Egypt, Iraq, Jordan, Lebanon, Tunisia, and to a lesser extent in Morocco and Pakistan (in annex 1A, refer to figure 1A.1). Business licensing burdens are also widespread and, for instance, cited as a "major" or "very severe" constraint by 35 percent of firms in Tunisia, 36 percent in Morocco, and 47 percent in Iraq (refer to annex 1A, figure 1A.1). Labor regulations present further hurdles.[5]

These business environment barriers create inefficiencies that hinder firms' ability to make productivity-enhancing investments such as engaging further in international trade and climate adaptation. As a result, most formal firms in the region remain small (40–75 percent have fewer than 20 employees, as shown in annex 1A, figure 1A.2), and in many countries, firms stay away from exporting: about 14 percent of firms export directly the equivalent of 7 percent of their sales (in annex 1A, refer to figure 1A.3).[6] In contrast, an analysis of World Bank Enterprise Survey (WBES) data shows that 18 percent of firms in Europe and Central Asia export about 9 percent of their sales.

SIGNIFICANT EFFECTS OF HEAT ON FIRMS' PERFORMANCE

Heat can have a negative impact on firms' productivity and jobs. These effects occur through various channels such as reduced input efficiency and higher operational costs as well as value chain supply and infrastructure disruptions (Costa et al. 2024; Goicoechea and Lang 2025; Grover and Kahn 2024; Kala et al. 2023). The theoretical framework for firms and climate analysis distinguishes between direct (internal to the firm) and indirect (external to the firm) impacts of hazards with a slow onset, such as rising temperatures (World Bank 2025), as shown in annex 1B, figure 1B.1.

In terms of direct impacts, high temperatures may impair workers' cognitive skills, increase errors and accidents, and lead to greater absenteeism and conflict in the workplace, which affects input efficiency and hence labor productivity. Rising temperatures also drive higher energy demand for cooling, thus raising operational costs and pressuring energy systems. As for indirect impacts, heat can also exacerbate power or water supply disruptions, which in turn are likely to increase production costs, or cause delays in deliveries which would affect productivity.

Increased Hazards and Exposure for Many Firms

Average temperatures in the region have increased since the 1950s (refer to annex 1C, figure 1C.1). In more recent years, comparing 1990–95 with 2018–23, average temperatures further rose across the region's countries (refer to annex 1C, figure 1C.2)—ranging from a 5 percent increase in Egypt to a 13 percent increase in Lebanon. Over the same time span, the average yearly number of "extreme heat" days has also risen in many countries—by up to 30 days in Egypt and 40 days in Jordan (refer to annex 1C, figure 1C.3).

In addition, clusters of firms in Egypt, Iraq, and Saudi Arabia have more annual days of extreme heat than the global average of 35 days per year (refer to map 1.1 and annex 1C, map 1C.1).[7] Although many economic hubs in the region are in cooler areas where the number of extreme heat days is relatively low (shaded in yellow in map 1.1), heat trends are still on the rise in these locations. For example, in Cairo, the average yearly number of extreme heat days went from 28 days in 1990–95 to 64 days in 2018–23, whereas in Tunis extreme heat days increased from about 4 to 11 days over the same time frame.[8]

A significant proportion of firms and jobs are thus exposed to extreme heat. Across the countries analyzed, nearly 40 percent of the firms (accounting for about 44 percent of jobs) face 50 or fewer extreme heat days per year. However, about 32 percent of firms (accounting for 21 percent of employment) are exposed 100 and more extreme heat days per year (refer to figure 1.1). Of the firms exposed to 100+ days, almost 60 percent are small (5–19 employees), with 16 percent being young (0–5 years).

MAP 1.1

MENAAP Countries Face a Large Number of Extreme Heat Days, 2023

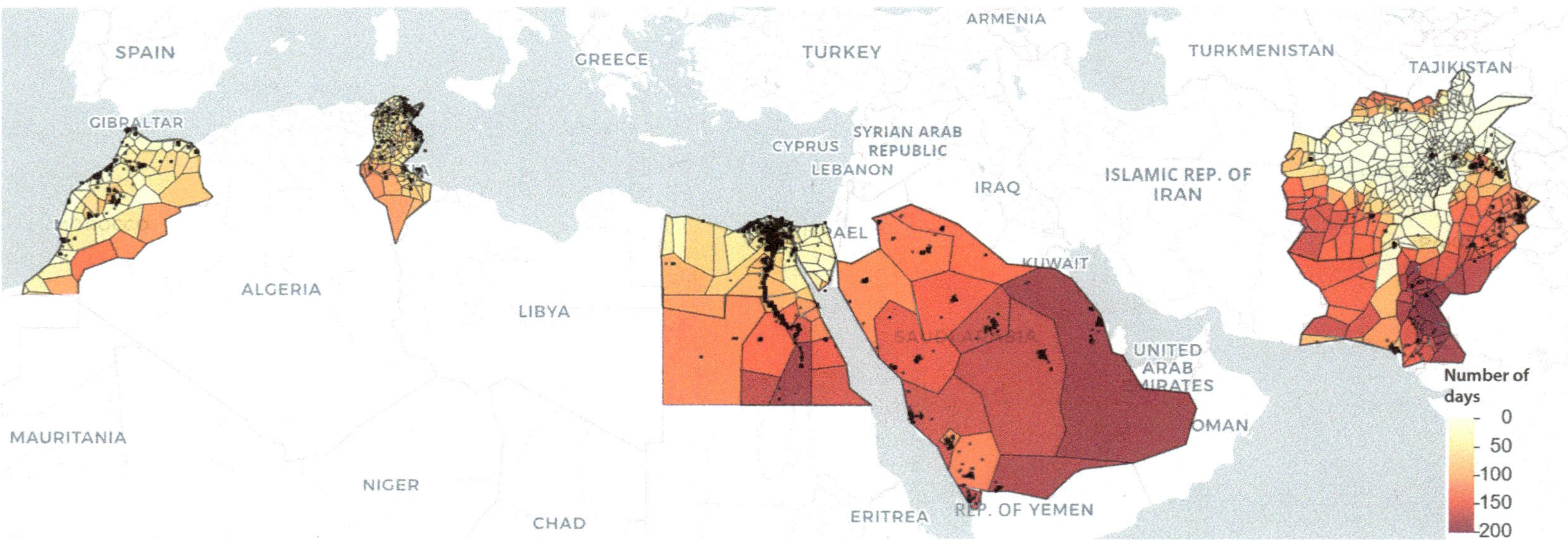

Source: Developed using the World Bank Firms and Jobs Vulnerability to Climate Change Explorer (https://datanalytics.worldbank.org/global-weather-map/).
Note: The heat map shows the number of "extreme heat" days per district, defined as days whose maximum temperature exceeds 35 degrees Celsius. The heat map's color coding is based on worldwide geocoded daily satellite weather data in 2023, indicating local exposure to heat day vulnerability. Dark red areas indicate the highest vulnerability worldwide. Black dots are locations of manufacturing and services firms from World Bank Enterprise Surveys. MENAAP = Middle East, North Africa, Afghanistan, and Pakistan.

FIGURE 1.1

An Important Proportion of Firms and Jobs Are in Locations with More Than 100 Extreme Heat Days Per Year

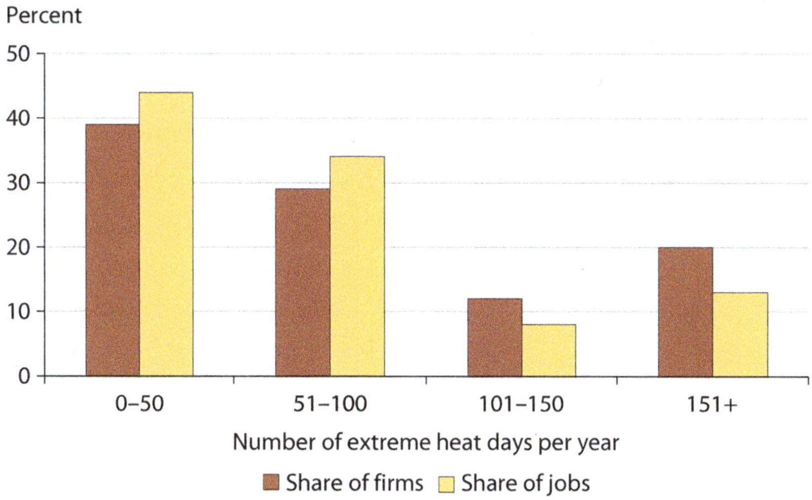

Sources: Calculations based on the World Bank Firms and Climate Adaptation Policy Toolkit (World Bank 2025) using World Bank Enterprise Surveys (WBES) and Database of Regional Climate Change data (Bettarelli et al. 2024).
Note: The figure shows the share of firms and firms' employment exposed to "extreme heat" days (whose maximum temperature exceeds 35 degrees Celsius. WBES data include Egypt, Arab Rep. 2020; Iraq 2022; Jordan 2024; Lebanon 2019; Morocco 2023; Saudi Arabia 2022; and Tunisia 2024.

Firm-Specific Vulnerability to Heat Events

Extreme heat affects jobs, input efficiency, and the operating costs of firms. A one standard deviation increase in the number of extreme heat days per year—equivalent to 17 additional days—has significant consequences for firm performance in MENAAP.[9] Firms experience notable declines in employment (2 percent) and wages (8 percent), which is a proxy for job quality, consistent with some of the CCDRs' findings (refer to figure 1.2, panel a).[10] They also face significant declines in factor efficiency: sales drop by 6 percent, sales per worker (a proxy for productivity) by 4 percent, and capital utilization by 3 percent.

For context, similar analysis on a global sample of firms finds that 10 additional extreme heat days reduce revenues by about 6–12 percent in low- and middle-income countries.[11] These results are consistent with heat-induced reductions in workers' productivity and disruptions to internal operations and align with key findings from CCDRs.[12]

There is also a positive association between extreme heat and the likelihood of investing in fixed assets (as discussed in annex 1E). One possible explanation for the latter is that higher temperatures may drive up repair and maintenance needs, prompting firms to invest in replacing or upgrading physical assets. Another is that heat-vulnerable firms are more likely to invest in adaptation practices. Indeed, there is a positive correlation between a firm investing in fixed assets and the likelihood that it adopts cooling systems, monitors water usage, and tracks emissions.[13] Similar results are found on the effects of average temperatures on input efficiency and operating costs, which are also discussed in annex 1E.

On the indirect side, extreme heat is associated with higher production costs by straining external systems: the probability of reporting insufficient water supply rises by 2 percent, highlighting that heat vulnerabilities in infrastructure are critical to firm operations.[14] These vulnerabilities carry a high economic price. It is estimated that water disruptions lead to utilization rate losses of US$6 billion a year in low- and middle-income countries (Hallegatte et al. 2019). Furthermore, the indirect losses from these water disruptions can exceed the direct losses by a ratio of 3 to 1, depending on the model specifications.

Small and older enterprises as well as firms not involved in global value chains (GVCs) are hit harder.[15] Small firms are disproportionately affected, with higher revenue and productivity losses resulting from increases in the number of extreme heat days. For instance, 17 more extreme heat days per year (that is, one standard deviation) reduces small firms' sales revenue by 10 percent and labor productivity by 9 percent. Older firms (over five years of age) are also more vulnerable to extreme heat, seeing a 7 percent drop in revenues and labor productivity losses of 5 percent.

FIGURE 1.2

Extreme Heat Strains Productivity, Jobs, and Production Costs, with the Sharpest Impact on Small Firms and Greater Resilience Among Young Firms and GVC Participants in MENAAP Countries, 2019–24

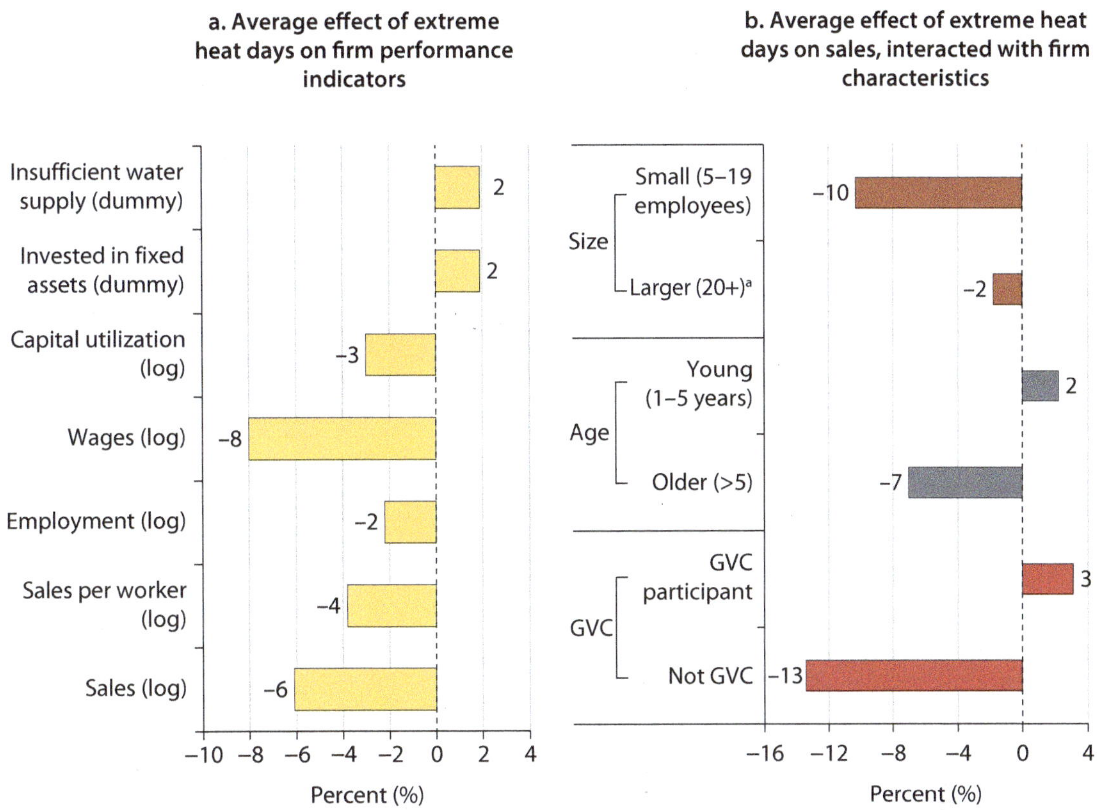

Sources: Calculations based on the World Bank Firms and Climate Adaptation Policy Toolkit 2025 using World Bank Enterprise Surveys (WBES) and Database of Regional Climate Change data (Bettarelli et al. 2024).
Note: "Extreme heat" days are those whose maximum temperature exceeds 35 degrees Celsius. Geocoded satellite weather data were matched with geocoded WBES data for over 10,000 firms across Egypt, Arab Rep.; Iraq; Jordan; Lebanon; Morocco; Saudi Arabia; and Tunisia between 2019 and 2024. Each bar corresponds to a separate regression. Coefficients are based on regressions on the effects of extreme heat days deviation from the long-term mean (independent variable) on performance (dependent variable), controlling for the long-term mean, firm size, and firm age and including country, sector, year, and country-sector fixed effects (for regressions, see annex 1E). Effects are in terms of one standard deviation increases in the number of extreme heat days—that is, 17 days in the Middle East, North Africa, Afghanistan, and Pakistan (MENAAP). Effects across performance variables are not independent and may overlap. The figure reports the combined coefficient when the dummy equals 1. For example, in panel b, the base coefficient of heat days on sales is −13 percent, but the interaction coefficient of heat days with being a global value chain (GVC) participant is 16.5 percent, resulting in a combined effect on sales of 3 percent for GVC participants. Results are robust to removing country-sector fixed effects and/or size and age controls. In panel b, the dotted line is not statistically significant. GVC participants are defined as firms that do any of the following: export, import, use foreign technology, or are foreign owned.
a. Results are statistically insignificant.

Young and large firms, in contrast, show greater resilience. Participation in GVCs also plays a role: firms that do not import or export, do not use foreign technology, and are not foreign owned suffer significantly greater sales and productivity losses in the face of extreme heat than firms with foreign links (refer to figure 1.2, panel b). Firm ownership is another important factor: foreign-owned firms and those that are part of a multiestablishment firm are more resilient.[16]

Heat shocks are not associated with an increase in electricity expenditure, possibly given a higher likelihood of power outages during these events. WBES data suggest that firms in areas experiencing an increased number of extreme heat days spend less on electricity than those in less affected areas. In the MENAAP region overall, for instance, 17 more extreme heat days per year relative to the long-run mean—that is, an increase by one standard deviation—decrease a firm's yearly electricity costs by 8 percent (refer to annex 1E, table 1E.3).[17] At the same time, an increase in the number of extreme heat days is associated with an increased use of generators to produce electricity (refer to figure 1.3, panel a)—a technology that can compensate for reduced electricity supply from the grid and generally runs on gas or other fuel (whose costs are not observed in the WBES data).

While not conclusive,[18] the evidence could suggest that firms in affected areas might cope with an increased risk of power outages by replacing the supply of electricity from the grid with an increased use of generators, a conclusion that is in line with anecdotal evidence (ICF 2024; Scott et al. 2014). More generally, WBES survey data confirm that the electricity supply remains comparatively unreliable in most of the countries in the region, a factor that can reduce firms' capacity to adapt to climate shocks and limit their competitiveness. Firms in most MENAAP countries report a substantially higher number of power outages than in aspirational peer countries such as southern European Union (EU) countries (refer to figure 1.3, panel b).

FIGURE 1.3

Extreme Heat Days Tend to Lower Firms' Direct Electricity Expenditure While Increasing the Use of Power Generators

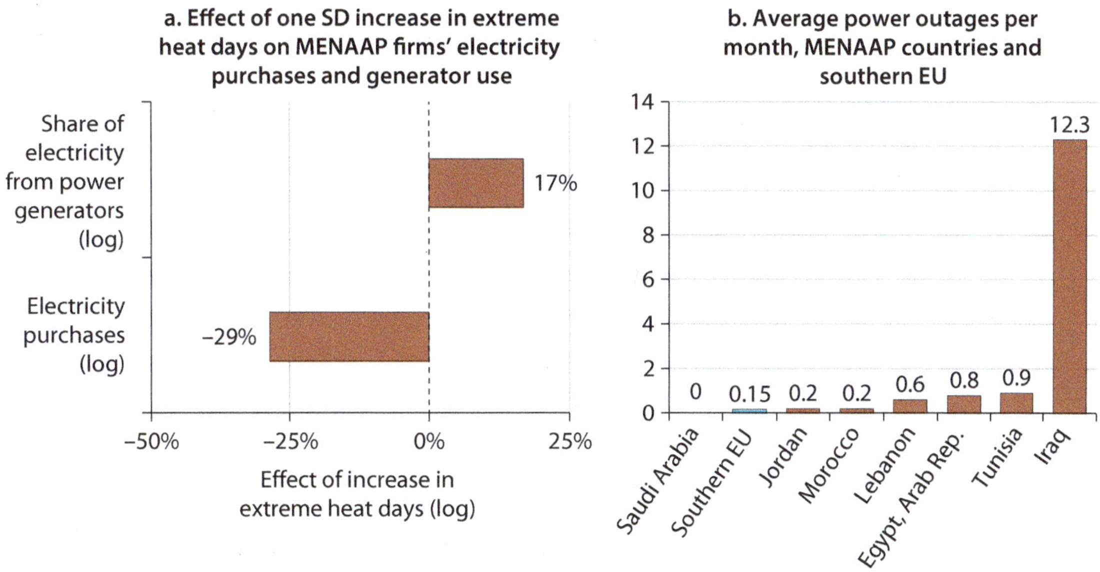

Sources: Calculations based on the World Bank Firms and Climate Adaptation Policy Toolkit (World Bank 2025) using World Bank Enterprise Surveys (WBES) and Database of Regional Climate Change data (Bettarelli et al. 2024).
Note: Extreme heat days are those whose maximum temperature exceeds 35 degrees Celsius. Coefficients based on regressions including heat days' deviation from the long-term mean, controlling for the long-term mean, firm size, and firm age, and including country, sector, year, and country-sector fixed effects. Effects are in terms of one standard deviation (SD) increases in the number of extreme heat days—that is, 17 days in MENAAP. Results are robust to removing country-sector fixed effects and/or size and age controls. The regression for "share of electricity from power generators" used a sample including Egypt, Arab Rep. 2020; Iraq 2022; Jordan 2019, 2024; Lebanon 2019; Morocco 2019, 2023; Saudi Arabia 2022; and Tunisia 2020, 2024. The regression for "electricity purchases" where a green module was available included Egypt, Arab Rep. 2020; Jordan 2019; Lebanon 2019; Morocco 2019; and Tunisia 2020. Monthly power outages are averaged from those reported by firms in WBES data, including Egypt, Arab Rep. 2020; Iraq 2022; Jordan 2024; Lebanon 2019; Morocco 2023; Saudi Arabia 2025, and Tunisia 2024. Southern European Union countries include France 2021, Italy 2024, Portugal 2023, and Spain 2024.

Limited Evidence of Firms' Adaptation Strategies

Despite evidence of the negative effects of heat, the take-up of adaptation practices among firms is still limited. Data show that firms that adopted heating or cooling improvements, implemented water management practices, invested in lighting systems, or set climate-related strategic objectives are significantly more resilient to the adverse effects of extreme heat days.[19] (See annex 1F for regressions interacting heat days and adaptation practices.)

However, few firms in the region are taking up these practices, and small firms are even less likely to do so—pointing to likely investment climate constraints. Egypt has the lowest overall take-up of cooling and water management adaptation strategies, with just 9 percent of firms investing in heating and cooling (refer to figure 1.4) and 3 percent of firms adopting water management practices (refer to annex 1F, figure 1F.1). Jordan, Lebanon, and Tunisia display the greatest differences in take-up between small and large firms. For example, 53 percent of large Lebanese firms are implementing water management practices, whereas only 6 percent of small firms are doing so (refer to annex 1F, figure 1F.1).

The analysis finds that constraints in access to finance play an important role in explaining the underinvestment in cooling. Indeed, for every unit of improvement in the access to finance index (a composite indicator that takes into account whether or not a firm has a savings and/or a checking account, an overdraft facility, and is an active borrower), the likelihood of investing in cooling increases by 5 percent (refer to annex 1F, table 1F.2). This finding is consistent with similar analysis in the South Asia region (Lang et al. 2025). Even in the presence of energy subsidies, firms may be discouraged from investing in cooling technologies because of high up-front costs and limited access to credit that delays or increases the cost of investment.

FIGURE 1.4

Few Firms Are Adopting Heating and Cooling Improvements

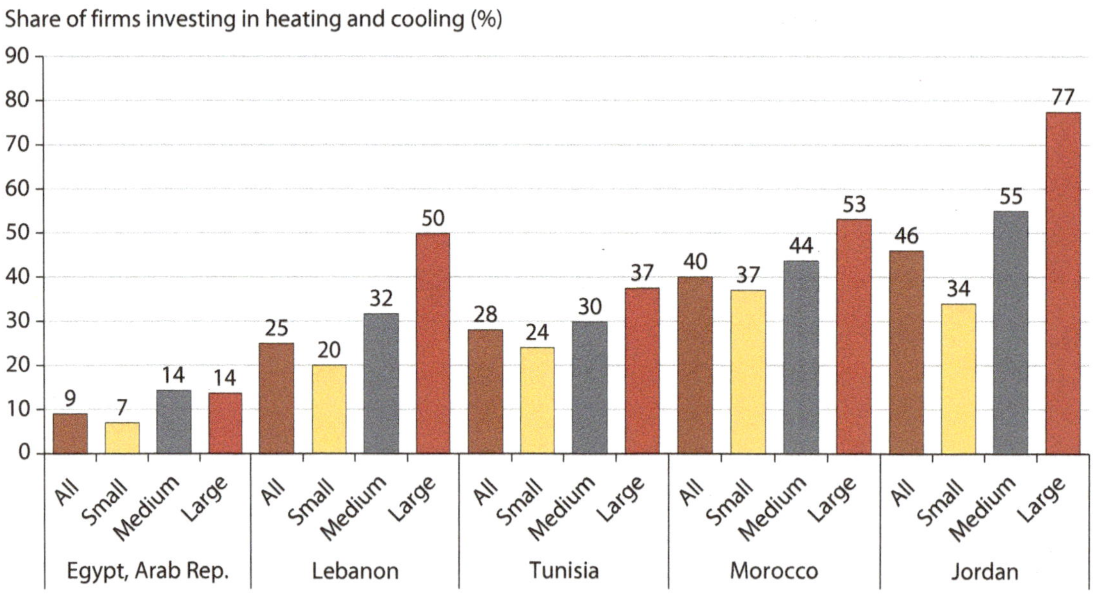

Source: World Bank Enterprise Surveys (WBES) data.
Note: The WBES data cover almost 6,000 firms across Egypt, Arab Rep. (2020); Jordan (2019); Lebanon (2020); Morocco (2020); and Tunisia (2020). Summary statistics computed using sampling weights. Small firms are those with 5–19 employees; medium firms, 20–99 employees; and large firms, 100 or more employees.

Energy Efficiency and Firm Characteristics

Countries in the region have room to improve the energy efficiency of their productive sector to cope with the costs of rising heat. Achieving higher energy efficiency—by lowering the energy consumed per unit of value added produced—is an important driver of competitiveness and an effective adaptation strategy.[20] It lowers production costs and reduces firms' dependence on power systems that are vulnerable to heat-related disruptions.

Economies in MENAAP tend to consume a comparatively high level of energy per unit of generated value added. Even after accounting for purchasing power differences, most MENAAP economies are more energy intensive than aspirational peers such as large southern European economies. For example, Tunisia and Saudi Arabia use, respectively, 34 percent and 101 percent more energy per unit of purchasing power parity (PPP) value added generated than the average of the five largest southern EU economies, according to International Energy Agency (IEA) data for 2021 (refer to figure 1.5, panel a).[21]

Energy efficiency can be improved by adopting practices and technologies already in use. Indeed, firm-level data show substantial heterogeneity in energy efficiency within the services and manufacturing sectors. The difference in energy efficiency (proxied by electricity cost as a share of sales) between the 25 percent least efficient firms and the 25 percent most efficient firms is large, ranging between a difference of 1.9 percentage points in Morocco to a difference of 3.9 percentage points in Iraq. These differences are nearly as large as the average firm's electricity intensity (the cost of electricity as a share of sales) in each country, equal to about 90 percent of the average in Morocco and 97.5 percent in Iraq (refer to figure 1.5, panel b).[22]

Energy efficiency is strongly linked to firm characteristics, with more productive, larger, and globally connected firms tending to be less energy intensive. Figure 1.6, panel a, shows that the 20 percent most productive firms (measured by sales per worker) are, on average, 61.3 percent more energy efficient than the 20 percent least productive, suggesting a complementarity between productivity and energy efficiency.

FIGURE 1.5

There Is Ample Room to Improve Energy Efficiency In MENAAP

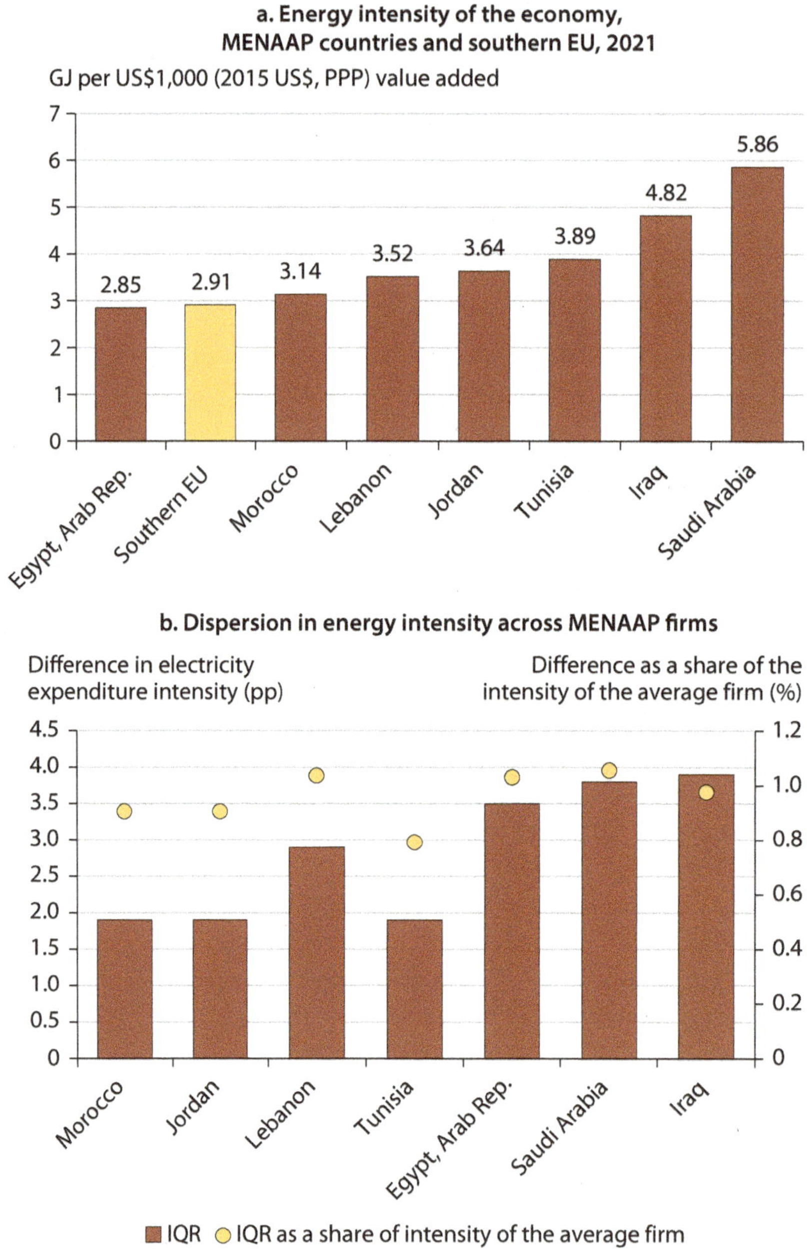

Sources: Calculations based on International Energy Agency estimates (panel a) and World Bank Enterprise Surveys (WBES) data (panel b).

Note: "Energy intensity" is indicated by gigajoules (GJ) of energy consumed per US$1,000 (2015 US$, purchasing power parity [PPP]) of value added in 2021. EU = European Union; MENAAP = Middle East, North Africa, Afghanistan, and Pakistan. WBES electricity cost and sales data are used to find the interquartile range (IQR)—the "difference" between the 25 percent least efficient firms and the 25 percent most efficient firms—of firms' electricity expenditure efficiency (intensity), shown by the bars (left axis), and the relative size of the IQR relative to the expenditure intensity of the average firm, shown by the diamonds (right axis). All measures control for sectoral averages. Surveys used: Egypt, Arab Rep. (2020); Iraq (2022); Jordan (2024); Lebanon (2019); Morocco (2023); Saudi Arabia (2022); and Tunisia (2024). MENAAP = Middle East, North Africa, Afghanistan, and Pakistan; pp = percentage points.

FIGURE 1.6

More Productive, Larger, and Exporting Firms Tend to Be More Energy Efficient

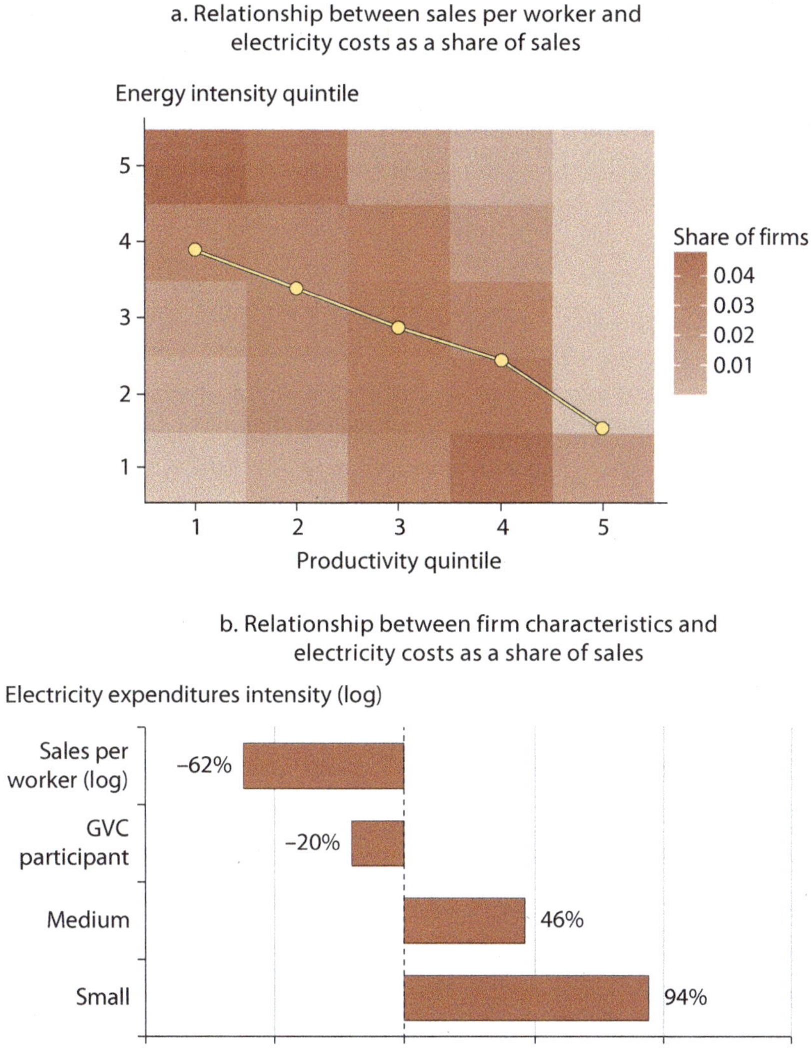

Source: Calculations based on World Bank Enterprise Surveys (WBES) data.

Note: "Energy intensity" is defined as a firm's yearly electricity costs divided by its yearly sales. WBES data are from Egypt, Arab Rep. (2020), Iraq (2022), Jordan (2019), Lebanon (2019), Morocco (2019, 2023), Saudi Arabia (2022), and Tunisia (2020). Panel b also includes Jordan (2024) and Tunisia (2024). In panel a, the orange circles plot the average quintile of energy intensity associated with the average firms in each productivity quintile ("productivity" measured by sales per worker). International Energy Agency (IEA) electricity price data were used to deflate electricity costs. In panel b, coefficients are drawn from individual regressions of energy intensity on the relevant variables. All regressions include country-sector and year fixed effects (FE) and firm size and age controls. Regressions are not weighted. Results are robust to removing country-sector FE and/or firm size and age controls. Small firms have 5–19 employees, medium firms, 20–99 employees, and large firms, 100 or more employees. Coefficients for medium and small firms correspond to effects relative to large firms.
GVC = global value chain.

Figure 1.6, panel b, shows that firms participating in GVCs are about 20 percent less energy intensive on average, even after accounting for size, likely reflecting the influence of competitive pressures and stricter standards. Larger firms also tend to be more energy efficient, pointing to technological or regulatory barriers that smaller firms face in adopting energy-saving practices. These findings underscore the potential for productivity-enhancing and energy-efficiency-increasing policies to reinforce each other and highlight the importance of easing constraints to boost the competitiveness of smaller firms.

TECHNOLOGY ADOPTION AS HEAT ADAPTATION

The adoption of "green" technology represents one of the most significant responses by firms to climate change challenges. Companies across sectors increasingly integrate sustainable innovations into their operations and products. Firms adopt these technologies to improve resource efficiency, reduce carbon footprints, comply with environmental regulations, and meet demands from more environmentally conscious consumers and investors. To complement the earlier analyses based on detailed climate and firm-level data, this section applies artificial intelligence (AI) methods to evaluate corporate communications data, tracking both the spread of green technologies and changing environmental regulatory pressures across markets.[23]

This report analyzes the adoption of green technology by reviewing earnings call transcripts (ECTs) from publicly traded companies for 2012–21 using natural language processing methods. ECTs are recordings of conversations between company executives and investors that discuss quarterly performance, strategic initiatives, and market challenges. Companies are classified as "green firms" when their ECTs mention technologies referenced in the World Intellectual Property Organization (WIPO) GREEN Database of Innovative Technologies and Needs.[24] This database comprises over 130,000 environmentally innovative technologies. The data are supplemented with firm characteristics and supply chain relationships to capture direct adoption and diffusion through business networks (Castro and Bastos 2025). For the MENAAP region, the analysis in this report covers 50 publicly traded firms and approximately 700 buyers and sellers.

Despite recent improvements in the greening of value chains and the speed of adoption of green technologies, MENAAP firms currently still lag behind in the use of green technologies. Figure 1.7, panel a, illustrates that approximately 14 percent of MENAAP firms mention green technologies in their ECTs, which is significantly lower than MENAAP's buyers and sellers (40 percent) and companies in other regions that are not buyers from and sellers to MENAAP firms (17 percent).

Nevertheless, the data also indicate that the proportion of "green firms" in MENAAP has increased from 0 percent in 2016 to 14 percent in 2021. This increase is aligned with a global surge in the proportion of "green" companies, especially since 2019. Morocco shows progress in emerging green technology implementation, though gaps remain between firms of different sizes and compared with high-income economies and other low- and middle-income countries (see annex 1H).

MENAAP firms, along with their buyers and sellers, also exhibit a growing emphasis on environmental regulations in their corporate communications. Here, environmental regulations in ECTs are detected by identifying sentences with regulatory terms and related words along with environmental keywords. Moreover, the ECTs are analyzed using the "GreenReg" Index,[25] which employs Meta's open-source Llama 3.2 large language model to assess regulatory pressure as increasing (+1), neutral (0), or decreasing (–1). This AI approach captures nuances and allows for accurate comparisons across regions

FIGURE 1.7

Mentions of Green Technologies and Environmental Regulations Are Rising

a. Percentage of firms mentioning green technologies in their ECTs by year

b. Percentage of firms mentioning environmental regulations in their ECTs by year

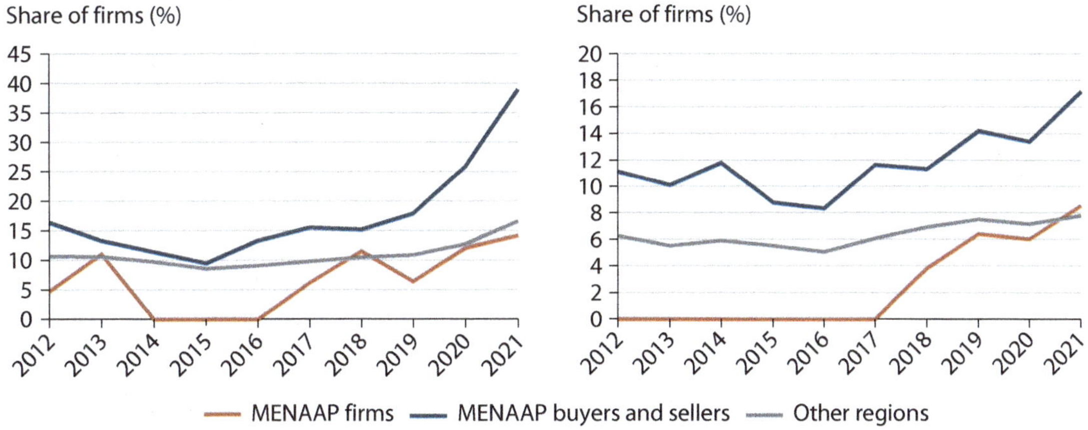

Sources: Calculations based on World Intellectual Property Organization (WIPO) and FactSet data.
Note: The sample included these Middle East, North Africa, Afghanistan, and Pakistan (MENAAP) countries: Bahrain; Egypt, Arab Rep.; Jordan; Kuwait; Morocco; Oman; Qatar; Saudi Arabia; and the United Arab Emirates. ECTs = earnings call transcripts.

and periods. It provides quantitative measures of evolving environmental regulatory pressures on firms (Bastos et al. 2024).

The share of MENAAP firms referring to green regulations is rising—from approximately 0.1 percent in 2017 to over 8 percent in 2021 (refer to figure 1.7, panel b). This proportion is notably higher among MENAAP buyers and sellers, reaching around 18 percent in the latter year, and exceeding other regions at approximately 8 percent. Globally, the mention of environmental regulations in boardrooms has noticeably increased since 2016–17, according to other calculations from the data.

MENAAP firms are under increasing regulatory pressure but still less than in other regions. The GreenReg Index shows whether firms perceive climate-related regulatory pressure to be rising or falling. Since 2019, MENAAP's green regulatory pressure has increased but is still lower than in the EU, the US, and the rest of the world. Globally, firms have reported an increase in environmental regulatory pressure over the past decade across various regions (refer to figure 1.8).[26]

FIGURE 1.8

MENAAP Firms Are Experiencing Increased Regulatory Pressure

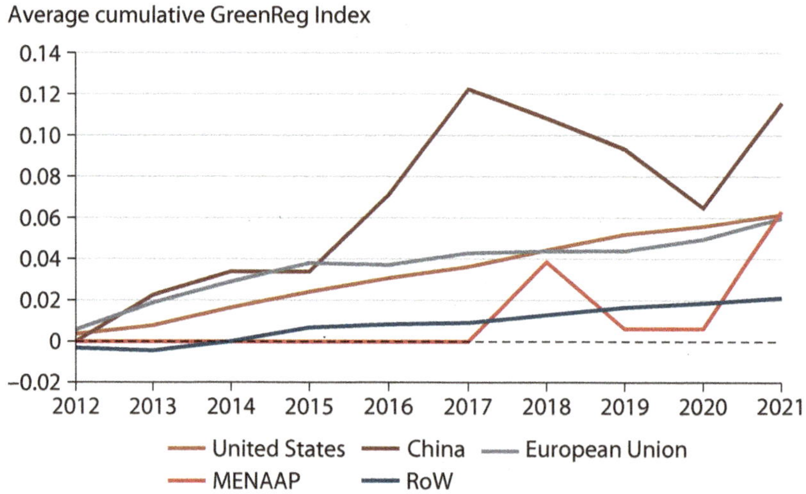

Sources: Calculations based on FactSet data and other official sources.
Note: The GreenReg Index (developed for this report based on Castro and Bastos 2025), using Meta's open-source Llama 3.2 large language model, analyzes earnings call transcripts to assess the extent to which regulatory pressure on firms is increasing (+1), neutral (0), or decreasing (−1). Middle East, North Africa, Afghanistan, and Pakistan (MENAAP) countries included in the sample are Bahrain; Egypt, Arab Rep.; Jordan; Kuwait; Morocco; Oman; Qatar; Saudi Arabia; and the United Arab Emirates. RoW=Rest of the World

HOW POLICIES CAN ALLEVIATE THE HARMFUL EFFECTS OF CLIMATE EVOLUTION ON FIRMS

Regulatory burdens and limited financial access exacerbate the negative impacts of heat on firms in MENAAP, suggesting that addressing the business environment is key for improving resilience. Firms faced with greater regulatory burdens (involving business licensing, labor regulations, and tax inspections) suffer more from the negative effects of extreme heat days on sales and productivity. Indeed, firms with the highest regulatory burden see a decline in sales of about 17 percent, while those facing median levels see a 5 percent drop as extreme heat days increase (annex 1I, figure 1I.1). Similarly, firms with limited access to bank accounts, credit, and/or overdraft facilities are also hit harder by costs associated with heat. Those facing the largest constraints in access to finance face a 21 percent decline in sales as heat days increase, while firms experiencing median levels see declines of 7 percent.

Evidence of a moderate relationship between MENAAP firms' productivity and adaptation investment suggests prioritizing business environment reforms that are tailored to the relevant climate risk and implemented in the most vulnerable locations.[27] Regressions of firm adaptation practices on labor productivity reveal a moderate, positive relationship between the two (refer to annex 1F, table 1F.1). Moreover, the likelihood of investing in cooling or heating for the most productive firms is positive, albeit less than 25 percent (annex 1F, figure 1F.6), and the likelihood of setting climate objectives is less than 15 percent (annex 1F, figure 1F.8).

Taken together, these results imply that policies that unlock firm productivity by addressing systemic business environment obstacles can in turn facilitate crucial adaptation investments—but such policies would benefit from focusing on the most vulnerable firms. For example, policy makers could consider streamlining licensing and permitting processes for heat-resilient buildings in affected areas, promoting the development of climate-related financial instruments, or introducing skills development programs for green jobs. These initiatives not only directly address heat-related issues but also can improve the business environment, boost firm productivity, and encourage further adaptation investments by firms themselves. Business environment reforms can be complemented with investments in resilient public infrastructure that benefit both the private sector and wider society.

NOTES

1. Average temperatures and extreme heat days are calculated using the World Bank's internal Climate Time Series Tool (https://datanalytics-int.worldbank.org/weather-timeseries/). Regional averages are composed of each MENAAP country's yearly value of the relevant indicator (except in West Bank and Gaza, for which data are unavailable).
2. Literature reviews by Goicoechea and Lang (2025) and Kala et al. (2023) summarize the literature along these key channels.
3. Unless otherwise specified, the numerical estimates and data in this chapter do not include Afghanistan and Pakistan.
4. Countries included are Djibouti, Egypt, Iraq, Jordan, Lebanon, Morocco, Saudi Arabia, Tunisia, and the Republic of Yemen.
5. Although limited to pre–COVID-19 World Bank Enterprise Surveys (WBES) rounds, World Bank (2022c) provides a detailed exploration of the top obstacles faced by firms in MENAAP.
6. These numbers do not include Afghanistan and Pakistan.
7. The global average of extreme heat days is calculated using the same climate data. This area-weighted average is based on second-level administrative divisions in geographic data (admin2-level) values for the number of heat days (whereby admin2 areas that are geographically larger are given a higher weight).
8. Whereas Tunisia as a country experienced an increase of 29 days (as shown in annex 1C, figure 1C.3), Tunis, the "economic hub" in the cooler area saw an increase of 7 days.
9. The analysis uses surveys of over 10,000 firms across Egypt, Iraq, Jordan, Lebanon, Morocco, Saudi Arabia, and Tunisia, covering 2019 to 2024. Note that the analysis focuses on past effects as opposed to predictions. Recent surveys are used because they were thought to more closely reflect current climatic conditions. The surveys also allow for a more accurate representation of the deviation between current temperatures or extreme heat days and their long-run means (calculated between 1980 and 2008). Regressions are presented in annex 1E.
10. For instance, Egypt's CCDR reports estimates from the International Labour Organization indicating that, for Africa as a whole, heat stress is projected to cause a 2.3 percent reduction in worked hours, the equivalent of 14 million full-time jobs (World Bank 2022a). Tunisia's CCDR finds that in a high-warming scenario, approximately 1,110 jobs in the tourism sector in coastal zones are expected to be lost by 2050 owing to sea level rise (World Bank 2023).
11. Revenue reduction estimates are based on the analysis of Berg et al. (2025). In this report, all countries except Saudi Arabia (high-income) and Iraq (upper-middle income) are lower-middle income.
12. For example, according to Egypt's CCDR, increasing temperatures from 26 to 31 degrees Celsius leads to an estimated 30 percent drop in labor productivity (World Bank 2022a). Lebanon's CCDR highlights that heat-affected labor productivity, alongside flooding and lower yields, is projected to reduce real GDP growth by 1.1 to 2.0 percent annually by 2040 (World Bank 2022b).

13. In terms of the direct relationship, there are statistically significant positive effects of increases in average temperatures on the probability of adopting adaptation measures. However, the effect of extreme heat days shocks is nonsignificant.
14. Firms are currently assessed in "isolation" because the WBES do not provide information on the identity or location of suppliers. However, it is important to consider ripple effects along supply chains; see Balboni et al. (2024) for evidence on the impact of extreme weather events on production networks. Results are thus likely to *underestimate* the impact of extreme heat on firm performance, as adverse effects propagated through supply chains are not accounted for. Measuring supply chain effects requires different microdata and methodologies outside the scope of this report.
15. Interacting heat days with firm characteristics highlights which traits are associated with better or worse performance under these extreme climatic conditions. For a full description of the variables used in the analysis, see annex 1D. Regressions of firm performance indicators on interactions between heat days and firm characteristics, along with country-sector and year fixed-effects and controls, are presented in annex 1E.
16. Although there are some positive effects of extreme heat on firm sales among younger and GVC-linked firms, these effects are small in magnitude. One possible explanation is that younger and GVC-linked firms are more likely to adapt, reducing their vulnerability. Indeed, being a GVC-linked firm is a strong predictor of take-up of adaptation practices (for example, investing in heating or cooling). These firms' relative resilience may lead to market reallocation, where more adaptable firms gain market share from those more adversely affected by heat, resulting in observed sales increases.
17. Introducing a squared heat days term in the regression reveals an inverted-U relationship: at lower levels of heat, electricity costs rise, but beyond a certain point, these costs decline (refer to annex 1E, table 1E.3). One possible explanation is that firms initially use more electricity for cooling, but as extreme heat days become more frequent, they may reduce production, shift activities, or even shut down, leading to lower energy consumption. Another is that firms may initially demand more electricity for cooling purposes, but as increasing levels of strain are put on the grid, power outages become more likely and thus electricity expenditures decline. There may also be country-specific factors at play: removing either Egypt or Saudi Arabia from the sample renders the effect of heat days on electricity cost insignificant. Results might be influenced by both countries having quite broad energy tariff structures, along with Saudi Arabia heavily subsidizing energy consumption.
18. The effect of extreme heat shocks on power outages is not significant (refer to annex 1E, table 1E.7).
19. The Green Economy module of the WBES grants insight into firms' energy efficiency and progress toward adapting to climate risks (resilience).
20. Higher energy efficiency is usually associated with higher productivity, as shown in Montalbano and Nenci (2019) and in figure 1.6, panel a, with evidence of a causal effect of energy efficiency on labor productivity (Kalantzis and Niczyporuk 2022).

21. The energy intensity indicator in figure 1.5, panel a, is taken from IEA estimates. It is equal to the amount of energy used by the economy per unit of value added generated, and it is meant to capture the efficiency at which the economy transforms energy sources into value added. This indicator is also used as the United Nations Sustainable Development Goals 7.3.1 indicator ("Energy intensity measured in terms of primary energy and GDP").
22. Electricity intensity is expressed as a share of sales (percent). The reported gaps (in percentage points) are absolute differences between high- and low-efficiency firms, which are then scaled by the average firm's electricity intensity to show their relative magnitude. For example, in Morocco, if the average firm spends 2.1 percent of sales on electricity, the 1.9 percentage point gap between efficient and inefficient firms equals approximately 90 percent of that average burden.
23. While the analysis of technology adoption focuses broadly on green technologies, future work could further narrow the scope to high potential cooling and efficiency technology in MENAAP.
24. Castro and Bastos (2025) show that references to green technologies in earnings calls are correlated with lower emissions at the firm level, particularly for large companies. This result suggests that mentions in earning calls serve as a proxy for green technology adoption.
25. The GreenReg Index was developed for this report by the World Bank (based on Castro and Bastos 2025).
26. Globally, perceived environmental regulatory pressure has increased across most climate-related areas—especially air, vehicles, energy, and general regulations (since 2012–14) and waste, water, and biodiversity (since 2016–17)—except in the agriculture sector, where it has declined since 2017.
27. An organizing principle to prioritize policies based on the degree of complementarity between adaptation investments and firm productivity is presented in the conceptual framework in annex 1B.

REFERENCES

Balboni, C., J. Boehm, and M. Waseem. 2024. "Firm Adaptation in Production Networks: Evidence from Extreme Weather Events in Pakistan." PEDL Working Paper, Growth Research Platform, Center for Economic Policy Research.

Bastos, P., L. Castro, and M. Garibotti. 2024. "Measuring the Costs of Environmental Regulations: A Deep Learning Approach." Unpublished manuscript, World Bank.

Berg, C., L. Bettarelli, D. Furceri et al. 2025. "Firm-Level Climate Change Adaptation: Micro Evidence from 134 Nations." Policy Research Working Paper 11081, World Bank.

Bettarelli, L., D. Furceri, M. Ganslmeier, and M. Schiffbauer. 2024. "The Database of Regional Climate Change (DRCC)." Unpublished manuscript, World Bank.

Castro, L., and P. Bastos. 2025. "The Emergence and Diffusion of Green Technologies: Firm-Level Evidence from Textual Analysis of Patents and Earnings Calls." Policy Research Working Paper 11036, World Bank.

Cirera, X., D. Comin, and M. Cruz. 2026. "Technology Sophistication Across Establishments." *Quarterly Journal of Economics*, ahead of print, March 23, 2026. doi:10.1093/qje/qjag018.

Costa, H., G. Franco, F. Unsal, S. Mudigonda, and M. Caldas. 2024. "The Heat Is On: Heat Stress, Productivity and Adaptation Among Firms." OECD Economics Department Working Paper 1828, Organisation for Economic Co-operation and Development.

Crick, F., M. Eskander, S. Fankhauser, and M. Diop. 2018. "How Do African SMEs Respond to Climate Risks? Evidence from Kenya and Senegal." *World Development* 108: 157–68. doi:10.1016/j.worlddev.2018.03.015.

De Haas, R., R. Martin, M. Muûls, and H. Schweiger. 2025. "Managerial and Financial Barriers to the Green Transition." *Management Science* 71 (4): 2890–921. doi:10.1287/mnsc.2023.00772.

Downey, M., N. Lind, and J. G. Shrader. 2023. "Adjusting to Rain Before It Falls." *Management Science* 69 (12): 7399–442. doi:10.1287/mnsc.2023.4697.

Gatti, R., A. Islam, C. Maue, and E. Zaveri. 2024. "Thirsty Business: A Global Analysis of Extreme Weather Shocks on Firms." Policy Research Working Paper 10923, World Bank.

Gatti, R., H. Onder, A. Islam, et al. 2025. *Shifting Gears: The Private Sector as an Engine for Economic Growth in the Middle East and North Africa*. MENA Economic Update, April, World Bank.

Gatti, R., Ozden, C., Bennet F. R., et al. 2026. *Challenges of Conflict and Industrial Policy for Development*. MENAAP Economic Update, April, World Bank.

Goicoechea, A., and M. Lang. 2025. "Firm-Level Adaptation to Climate Change in Low- and Middle-Income Countries." *The World Bank Research Observer* 41 (1): 157–89. doi:10.1093/wbro/lkaf001.

Grover, A., and M. E. Kahn. 2024. "Firm Adaptation to Climate Risk in the Developing World." Policy Research Working Paper 10797, World Bank.

Hallegatte, S., J. Rentschler, and J. Rozenberg. 2019. *Lifelines: The Resilient Infrastructure Opportunity*. Sustainable Infrastructure Series. World Bank.

Heger, M., L. Vashold, A. Palacios, M. Alahmadi, M.-A. Bromhead, and M. Acerbi. 2022. *Blue Skies, Blue Seas: Air Pollution, Marine Plastics, and Coastal Erosion in the Middle East and North Africa*. World Bank.

ICF. 2024. "Extreme Heat in Disadvantaged Communities." Report, ICF Climate Center, ICF International.

Islam, A., D. Moosa, and F. Saliola. 2022. *Jobs Undone: Reshaping the Role of Governments Toward Markets and Workers in the Middle East and North Africa*. Middle East and North Africa Development Report. World Bank.

Islam, A., D. Moosa, and F. Saliola. 2023. "Are MENA Businesses Ready for a Green Transition?" *Perspectives: MENA, Afghanistan, and Pakistan* (blog), March 16, 2023. https://blogs.worldbank.org/en/arabvoices/are-mena-businesses-ready-green-transition.

Kala, N., C. Balboni, and S. Bhogale. 2023. "Climate Adaptation." *VoxDevLit* 7 (1).

Kalantzis, F., and H. Niczyporuk. 2022. "Labour Productivity Improvements from Energy Efficiency Investments: The Experience of European Firms." *Energy* 252: 123878.

Lang, M., J. Rexer, S. Sharma, and M. Triyana, eds. 2025. *From Risk to Resilience: Helping People and Firms Adapt in South Asia.* Conference Edition. World Bank.

Lopez-Acevedo, G., M. Ranzani, N. Sinha, and A. Elsheikhi. 2023. *Informality and Inclusive Growth in the Middle East and North Africa.* Middle East and North Africa Development Report. World Bank. doi:10.1596/978-1-4648-1988-9.

Montalbano, P., and S. Nenci. 2019. "Energy Efficiency, Productivity and Exporting: Firm-Level Evidence in Latin America." *Energy Economics* 79: 97–110.

Scott, A., E. Darko, A. Lemma, and J-P. Rud. 2014. "How Does Electricity Insecurity Affect Businesses in Low- and Middle-Income Countries?" Briefing 01, Overseas Development Institute.

Seneviratne, S. I., X. Zhang, M. Adnan, et al. 2021. "Weather and Climate Extreme Events in a Changing Climate." In *Climate Change 2021: The Physical Science Basis. Working Group I Contribution to the Sixth Assessment Report of the Intergovernmental Panel on Climate Change*, edited by V. Masson-Delmotte, P. Zhai, A. Pirani, et al., 1513–1766. Cambridge University Press. doi:10.1017/9781009157896.013.

WMO (World Meteorological Organization). 2026. "WMO Confirms 2025 Was One of the Warmest Years on Record." Press release, January 14, WMO. https://wmo.int/news/media-centre/wmo-confirms-2025-was-one-of-warmest-years-record.

World Bank. 2022a. "Egypt—Country Climate and Development Report." Report 177494, World Bank.

World Bank. 2022b. "Lebanon—Country Climate and Development Report." Report 188143, World Bank.

World Bank. 2022c. *Unlocking Sustainable Private Sector Growth in the Middle East and North Africa: Evidence from the Enterprise Survey*. European Investment Bank, European Bank for Reconstruction and Development, and World Bank.

World Bank. 2023. "Tunisia—Country Climate and Development Report." Report 186032, World Bank.

World Bank. 2025. "Firms and Climate Adaptation Policy Toolkit Guidance Note." Unpublished manuscript, World Bank.

CHAPTER 2

Impacts of European Climate Mitigation Policies on MENAAP's Trade and Foreign Direct Investment

INTRODUCTION

European countries are intensifying efforts to combat climate change, which is of particular interest to Middle East, North Africa, Afghanistan, and Pakistan (MENAAP) countries given their close trade ties. Two key European Union (EU) climate mitigation policies include the Carbon Border Adjustment Mechanism (CBAM) and the Corporate Sustainability Due Diligence Directive (CS3D). The EU CBAM has influenced other economies, such as the United Kingdom, to consider their own versions of similar mechanisms. These policies, which aim to reduce carbon emissions and encourage a green transition, are expected to reshape trade and foreign direct investment (FDI) in the MENAAP region.

MENAAP's trade and FDI inflows will be affected by European policies given the region's heavy dependence on carbon-intensive exports. MENAAP exports 6.7 percent of global aluminum, cement, fertilizer, and iron and steel, despite accounting for only 0.06 percent of global trade in 2022. Over one-sixth of the region's carbon-intensive exports are directed to the EU (16 percent) and the UK (1.7 percent) and thus are affected by their CBAMs.[1]

MENAAP firms integrated into EU value chains for goods like textiles, agriculture, and raw materials are also affected by the EU CS3D (discussed in box 2.2), which sets requirements on firms that are part of EU value chains. These sectors now face stricter sustainability and human rights due diligence requirements, which could constrain MENAAP firms' access to EU markets.

These European policies will also affect FDI inflows into MENAAP. Higher export costs to the EU and United Kingdom (UK) may affect FDI in various ways. FDI to countries with high-emission heavy industries like steel and aluminum might decrease because of the larger capital investment needed to decarbonize these sectors. Conversely, FDI might increase in economies with energy-efficient heavy industries with lower decarbonization requirements.

This chapter is structured as follows: The next section describes the provisions of the EU CBAM. Later sections assess MENAAP countries' exposure to the policy and estimate the impact of the policy on MENAAP countries' sectors and economies. A similar exposure approach is followed for the UK CBAM, which is presented in annexes 2A and 2B.

THE EU CARBON BORDER ADJUSTMENT MECHANISM

The EU CBAM is a policy designed to reduce greenhouse gas emissions by accounting for the carbon costs of imported goods. Its goal is to prevent carbon "leakage," a potential consequence of the EU Emissions Trading System (ETS), whereby industries with high emissions relocate production to regions with less stringent climate policies or substitute domestic production with imports of high-carbon goods. By requiring EU importers to purchase CBAM certificates for carbon dioxide (CO_2) emissions not priced domestically, the mechanism aims to mirror the EU ETS for foreign producers, leveling the playing field for businesses and creating incentives for non-EU companies to reduce emissions.

Adopted in April 2023, CBAM will be phased in from 2026 to 2034. Initially, it will cover high-emission sectors like aluminum, cement, electricity, fertilizer, hydrogen, and iron and steel, as well as certain precursors (such as cathode active materials) along with limited downstream products (such as screws and bolts). It will cover direct emissions (scope 1) for aluminum and for iron and steel. It will cover both direct (scope 1) and indirect emissions (scope 2) for other sectors (refer to figure 2.1).[2] Its coverage is expected to expand over time, reflecting the ETS's evolving coverage. EU decisions on the speed and coverage of such expansion will be informed by periodic evaluations.

As of October 1, 2023, CBAM requires EU importers to report embedded emissions, with payments starting on January 1, 2026. From that date, imports of CBAM goods into the EU will be restricted to preauthorized CBAM declarants (the importers), who must pay a carbon price similar to that paid by EU companies under the EU ETS. Authorized CBAM declarants must declare the previous year's embedded emissions in imported goods (using actual or default emissions intensities), surrendering corresponding CBAM certificates. Importers must undergo independent verification of reported emissions and continuously monitor emissions levels. Any direct carbon price paid in the exporting country will be deducted from the CBAM fiscal obligation.[3]

FIGURE 2.1

Adopted in 2023, CBAM will be Phased In from 2026 to 2034

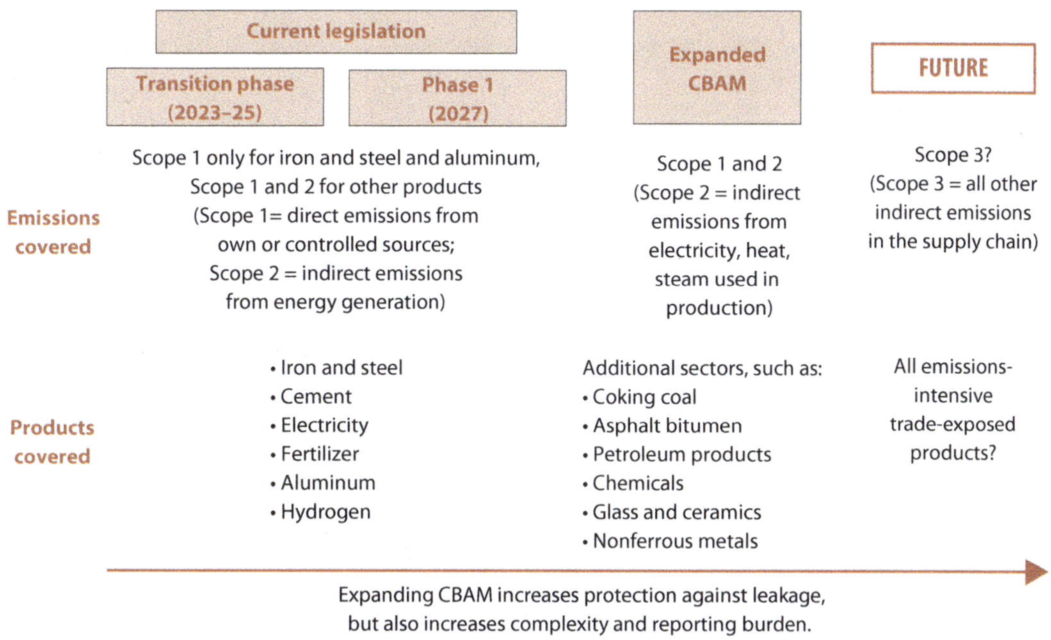

Source: World Bank elaboration based on European Commission 2025b.
Note: CBAM = Carbon Border Adjustment Mechanism.

Exporters unable to verify carbon emissions may face penalties starting in 2026. When actual emissions cannot be reliably verified, the average emission intensity of the exporting country, increased by a proportionately designed markup, will be applied as the default value. These default values are yet to be decided by the European Commission through an implementing act. This may disadvantage producers from countries that are able to manufacture CBAM products with embedded emissions lower than these default values. The markup encourages producers and importers to use verified actual emissions data.

By 2034, the EU's CBAM will require both EU producers and non-EU exporters to fully pay for carbon emissions, as free allowances to EU producers will be phased out to promote emissions reductions. Currently, EU producers receive free allowances for emission permits to help them compete with foreign competitors. This system levels the playing field by offsetting carbon costs for domestic industries. By 2034, the proportion of free allowances will have been gradually phased out, fully transferring the carbon cost burden onto producers and encouraging emissions reductions. At the same time, because CBAM will be phased in gradually alongside the phaseout of free allowances under the EU ETS, the fiscal obligation for non-EU producers will start lower and progressively increase until CBAM is fully implemented in 2034.[4] At that point, both EU producers and non-EU exporters of CBAM products will be required to pay equally for carbon emissions, because no free allowances will be available to EU producers.

In February 2025, the European Commission proposed amendments to the EU CBAM to ease administrative burdens, effective from January 1, 2026. The EU legislative bodies (Council of the EU and the European Parliament) reached an agreement on these amendments in June 2025, and the final legal act was decided in October 2025. The major modifications include the following:

- *Scope and threshold for application of CBAM.* The threshold will be based on an annual cumulative net mass per importer and set at 50 tons, replacing the original threshold of €150 for each consignment of imported goods. This change excludes importers of small quantities of goods, still ensuring that more than 99 percent of emissions are maintained in scope.
- *Carbon price paid.* CBAM declarants (importers) may use default carbon prices calculated and to be provided by the European Commission by an upcoming implementing act. Although the methodology for calculating these country-specific carbon prices has not yet been

specified at the time this report was written, previous EU legislation has referenced an "effective carbon price paid," which accounts for any rebates or reduction mechanisms weakening the carbon price paid. Where the embedded emissions are determined on the basis of default values, a reduction in the CBAM financial obligation may be claimed only by a reference to these yearly default carbon prices.

- *Date to submit declaration.* The submission deadline for CBAM declarants is shifted by four months, from originally May 31 the year following importation to September 30 of that year. Although this new date may provide additional time to gather necessary information and ensure the verification of embedded emissions, this extension does not reduce the overall methodological complexity of reporting and verification.
- *Default emission intensity.* The current CBAM design empowers the European Commission to set default values using the best available data for each exporting country and type of goods. If reliable emission data for the exporting country are not available, the amendment obliges the Commission to use the country-specific default and any other countries' default emission intensity from the Annexes of CBAM Implementing Regulation (EC 2025a).[5]

The impact of the EU CBAM on the competitiveness of an exporter in the EU market from a country without carbon pricing depends on three factors:

- The higher its CO_2 emissions intensity compared with the average EU firm, the greater the impact due to increased carbon payments.
- A higher concentration of exports to the EU will amplify the impact on the exporter.
- The administrative and compliance costs related to calculating and reporting embedded emissions at the product level can further increase the burden on exporters, especially smaller firms with limited capacity to meet reporting requirements.

Although CBAM is applied at the firm level, this report evaluates the exposure to CBAM at the sector level given the lack of relevant firm-level data.[6]

CBAM Trade Exposure Index

The exposure of a country's affected sector to the EU CBAM is assessed based on the CBAM Trade Exposure Index (TEI), which captures the first two determinants of the impact of the EU CBAM: relative emission intensity and export dependence (Maliszewska et al. 2025). The index is formulated as the product of the sector's emission intensity relative to average EU competitors and its export dependence on the EU as a destination market, as follows (equation 2.1):

$$\text{CBAM TEI}_{c,s} = \frac{\overbrace{\tau.(EI_{c,s} - EI_{EU,s}) \cdot Xq_{c,s}^{EU}}^{\text{excess carbon payments}}}{X_{c,s}^{EU}} \cdot \frac{X_{c,s}^{EU}}{\underbrace{X_{c,s}^{World}}_{\text{Total trade in the sector}}} \tag{2.1}$$

where τ = CO_2 price, c = country, s = sector, X = exports in value, Xq = exports in quantity, and EI = emission intensity (ton of CO_2 equivalent per ton of export).

This index is equivalent to dividing the excess carbon payments—determined by the carbon price per ton of CO_2 equivalent (tCO_2e), the quantity of the sector's affected exports to the EU, and the difference between a country's emissions intensity and the EU average in the affected sector—by the total value of the country's global exports of the affected sector.[7] Excess carbon payments are the total value of CBAM certificates in excess of the emissions costs of EU competitors for the quantity of export to the EU of the affected sector. A high positive index indicates that exporting firms in a country's affected sector will incur higher carbon payments than the average EU firm for the same products, which could reduce their competitiveness in the EU market and decrease exports of the sector. Conversely, a negative index suggests that exporting firms in a country's affected sector will incur lower carbon payments than the average EU firm, which could enhance their competitiveness in the EU market and increase exports of the sector.

To compute this index, we extract trade data from the International Trade Analysis Database (BACI) and emission intensity data from the European Commission (EC 2025a; Vidovic et al. 2023).[8] Export data are taken from BACI as mirror values, hence reflecting the value of imports into the EU as reported by the EU. We assume that the CBAM financial obligations will be ultimately paid by exporters, even if the legislation requires EU importers to surrender the CBAM certificates. We also value the excess carbon payments for 2034 onward once the CBAM is fully phased in.

The carbon price is set at US$100 per tCO_2e, assuming the prevailing price will reach this amount by 2034 when CBAM is fully implemented. The carbon price aligns with European Union Allowance futures trading.

Emission intensities can fluctuate over time, vary across firms, and depend on data collection methods and sample coverage, which may influence the accuracy and interpretation of competitiveness outcomes. We employ data from the latest default emission intensities published by the European Commission (EC 2025a). In the case of MENAAP, the database provides emission data only for the aluminum, cement, fertilizer, and iron and steel sectors. This analysis therefore does not cover the other CBAM sectors: hydrogen production and electricity.

CBAM Certificate Payments, by Country and Sector

The total CBAM certificate payments made by MENAAP countries and EU producers differ depending on the sector. These payments are determined by multiplying emission intensity, export volume to the EU, and the carbon price. The ad-valorem equivalent (AVE) of CBAM certificate costs for both MENAAP economies and EU producers is shown as a percentage of exports to the EU (figure 2.2, panels a, d, g, and j) and can be understood as a tariff equivalent of CBAM certificate costs.[9]

The difference in total CBAM certificate payments between countries and EU producers, or difference in carbon payments, is referred to as "excess carbon payments." The EU AVE differs by country because of the varying subproduct composition of that sector's exports across countries. In the case of aluminum, the AVE for the Arab Republic of Egypt is 8.5 percent, meaning that the CBAM certificate payments amount to 8.5 percent of the value of Egypt's aluminum exports to the EU, which is 1.2 times higher than the EU producers' AVE. Meanwhile, Morocco's AVE for aluminum is 1.4 percent, which is about one-third of the EU producers' AVE. This means Egyptian aluminum exporters will face higher CBAM certificate costs (positive excess carbon payments) than the EU producers, whereas Moroccan aluminum exporters will face lower CBAM certificate costs than the EU producers (negative excess carbon payments).

Though CBAM will likely have a small impact on MENAAP economies, the CBAM TEI indicates that CBAM may pose challenges for specific sectors in certain MENAAP countries. The index values for aluminum, cement, fertilizer, and iron and steel across MENAAP countries are presented in panels c, f, i, and l, respectively, of figure 2.2. These graphs present the CBAM TEI capturing excess carbon payments (that is, expected default payments in excess of EU averages for the product mix of the exporting country) as a share of global exports of the sector. The graphs in panels b, e, h, and k illustrate the two components of the CBAM TEI—that is, the trade exposure and excess carbon payments for each county as a share of export to the EU.

FIGURE 2.2

CBAM Trade Exposure Index and Key Components of MENAAP Countries in Main EU CBAM Sectors

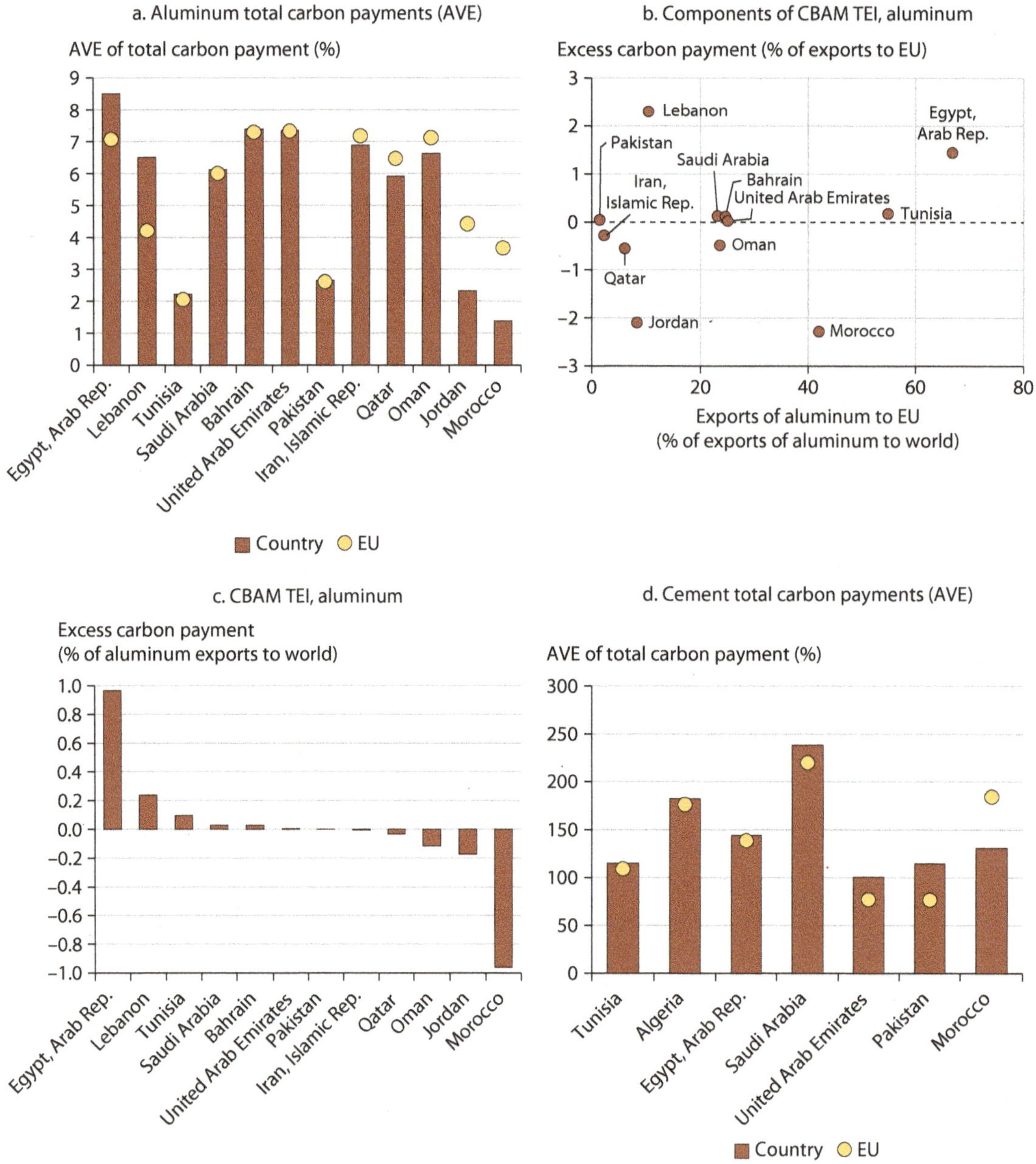

(continued)

FIGURE 2.2

CBAM Trade Exposure Index and Key Components of MENAAP Countries in Main EU CBAM Sectors (*continued*)

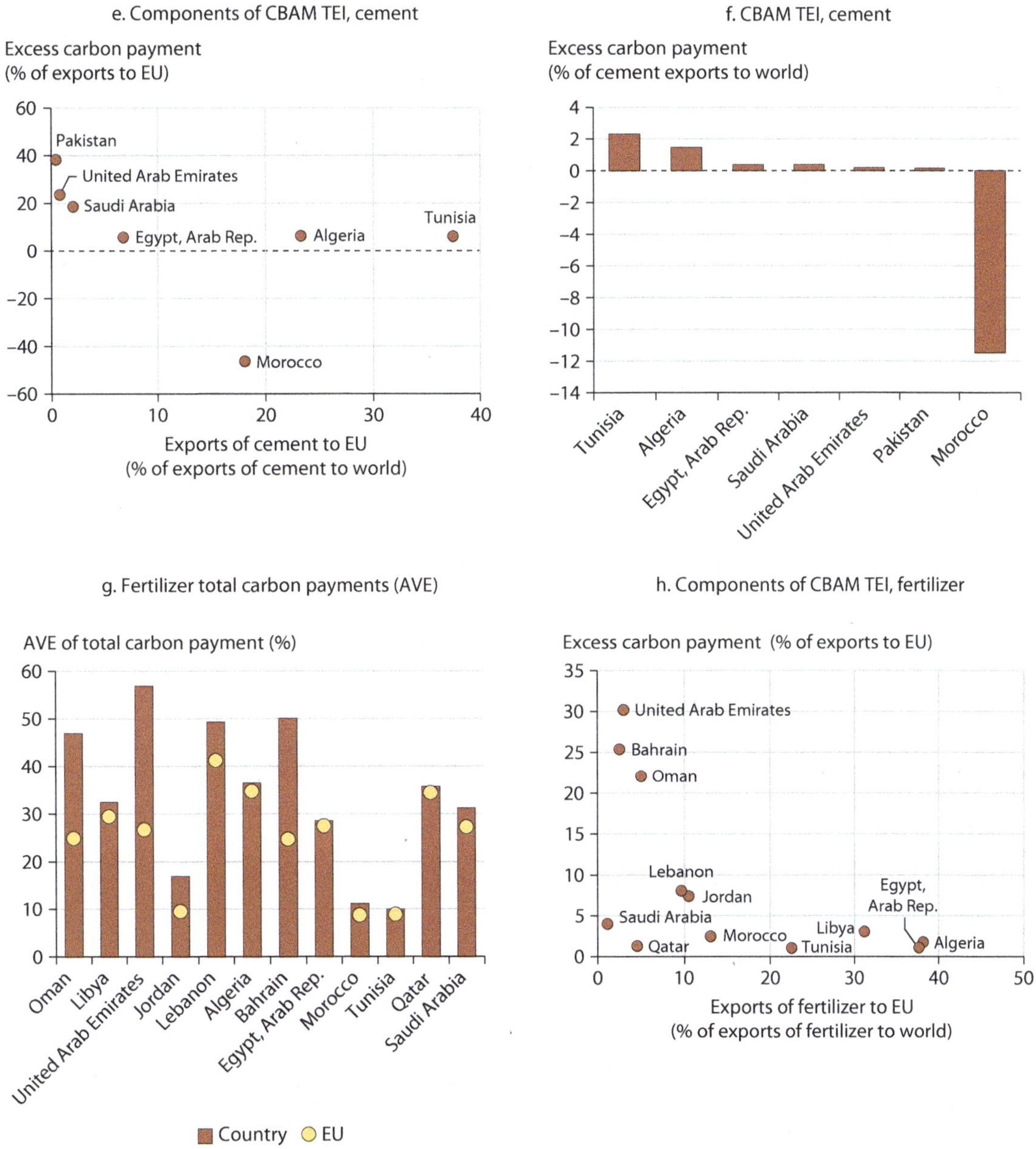

(continued)

FIGURE 2.2

CBAM Trade Exposure Index and Key Components of MENAAP Countries in Main EU CBAM Sectors (*continued*)

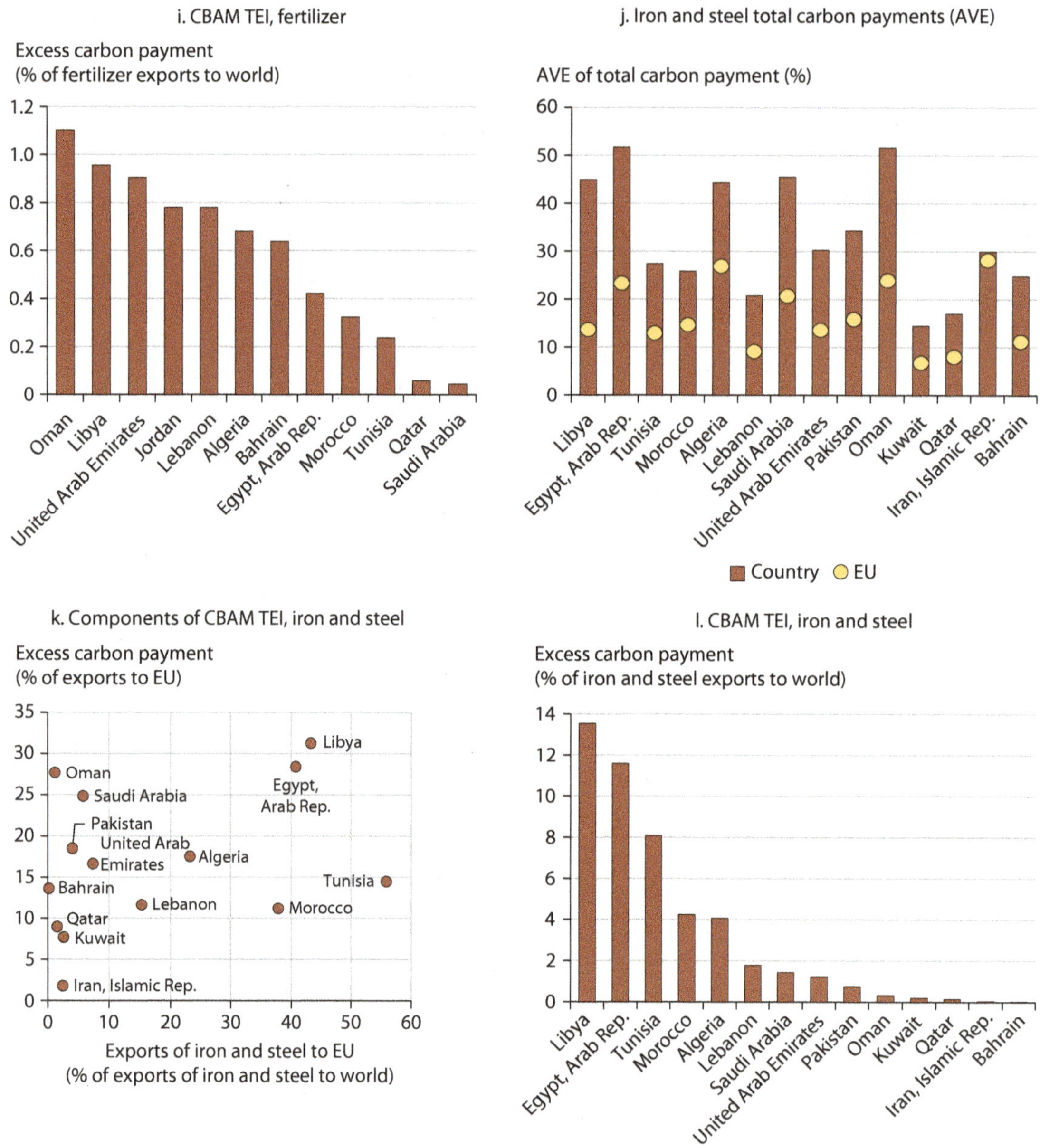

(continued)

FIGURE 2.2

CBAM Trade Exposure Index and Key Components of MENAAP Countries in Main EU CBAM Sectors (*continued*)

Sources: World Bank calculations using Maliszewska et al. 2025 based on 2023 trade data from the International Trade Analysis Database (BACI) and default emission intensities from EC 2025a.
Note: Panels a, d, g, j represent ad valorem equivalent (AVE) of total embedded carbon payments as a share of total exports of these products to the EU. To compare to the embedded carbon payments that an EU producer would have for the same mix of products, the EU average emissions intensity for each product is multiplied by the country's exports by weight to the EU, and aggregated across the relevant products in the sector. Panels b, e, h, and k present the trade exposure and excess carbon payment in the aluminum, cement, fertilizer, and iron and steel sectors, respectively. They present the decomposition of the Trade Exposure Index (TEI) into two components: carbon intensity relative to the EU and export dependence. Panels c, f, i, and l express the Carbon Border Adjustment Mechanism (CBAM) TEI in percentage terms, capturing the excess embodied carbon payments as a share of the value of the country's global exports of the affected sector (to achieve this, the TEI value is multiplied by 100). The carbon price is set at US$100 per metric ton of carbon dioxide (CO_2) equivalent— which is the assumed value that the carbon price will reach by 2034, when the CBAM is fully phased in. Countries with an export value to the EU less than US$1 million are excluded. AVE = ad-valorem equivalent; EU = European Union; MENAAP = Middle East, North Africa, Afghanistan, and Pakistan.

The CBAM TEI suggests that several MENAAP countries may lose competitiveness in aluminum, cement, fertilizer, and iron and steel.[10] Of those MENAAP countries for which there is default emission intensity data, the most exposed are Egypt in aluminum, Tunisia in cement, Oman in fertilizer, and Libya in iron and steel, as follows:

- *Egypt is the most exposed in aluminum* (refer to figure 2.2, panel c), as 67 percent of its aluminum exports go to the EU, with the expected excess carbon payment of 1.4 percent of the value of its aluminum exports going to the EU (refer to figure 2.2, panel b).
- *Tunisia is the most exposed in cement* (refer to figure 2.2, panel f), as 38 percent of Tunisia's cement exports are destined for the EU, and its exporters face an excess carbon payment of 6.1 percent of those exports (refer to figure 2.2, panel e).
- *Oman is the most exposed in fertilizer* (refer to figure 2.2, panel i), although 5 percent of its fertilizer exports are shipped to the EU and would see an excess carbon payment of 22 percent by value (refer to figure 2.2, panel h).
- *Libya is the most exposed in iron and steel* (refer to figure 2.2, panel l), as 43 percent of its iron and steel exports go to the EU, and they face an excess carbon payment of 31 percent (refer to figure 2.2, panel k). Furthermore, its emissions intensity is high relative to EU counterparts, leading to the highest TEI rate across all sectors.

Higher costs from CBAM certificates may reduce competitiveness on EU markets, potentially prompting industries to seek energy-efficient technologies to decrease carbon intensity or to redirect exports to other markets.

In contrast, some MENAAP countries like Morocco have a negative CBAM TEI in aluminum and cement and could gain competitiveness in that sector. A negative CBAM TEI implies that the cement products that are most heavily traded by Morocco are less carbon intensive than the EU emission intensities for those products, signaling that exporters may benefit from greater competitiveness in the EU market. Although Morocco's total carbon payments for its cement exports to the EU represent a high proportion of its export values—with AVEs at approximately 131 percent (refer to figure 2.2, panel d)—the default emission values for Moroccan cement products exported to the EU are actually lower than default values applied to EU producers, resulting in a negative CBAM TEI (refer to figure 2.2, panel f).

CGE MODELING OF CBAM IMPACTS

After reviewing the exposure of MENAAP's aluminum, cement, fertilizer, and iron and steel sectors to the EU CBAM, we now assess the impacts of the EU CBAM using computable general equilibrium (CGE) modeling.[11] The exposure analysis focuses on the supply side, identifying sectors likely to be affected by CBAM. A comparative static Global Trade Analysis Project (Energy) (GTAP-E) CGE model enables us to assess the general equilibrium impact capturing interactions between all markets simultaneously—goods, services, and factors of production—as it incorporates input-output linkages and economywide effects. As a result, the CGE analysis can assess the relative competitiveness of MENAAP countries on the EU market with respect to EU domestic producers and other exporters.

The analysis considers supply and demand changes resulting from the removal of free allocations in the EU and the imposition of CBAM. The model incorporates a GTAP 11 Circular Economy Data Base (GTAP11-CE), which includes 99 sectors with detailed representations of primary, secondary, and recycling activities for metals and plastics.[12] To depict a stylized pre-CBAM scenario, we establish a baseline with a uniform carbon price of US$100 per ton of CO_2 across all sectors in the EU, maintaining 90 percent free allocations in the aluminum, cement, fertilizer, and iron and steel sectors.[13] The CBAM scenario, on the other hand, maintains the same carbon price but assumes the complete removal of free allocations in the EU and imposes financial obligations on imports of CBAM products, representing the full implementation of CBAM.

The analysis employs the latest data on emission intensities, as in the exposure analysis. We take the trade weighted average emission intensity of the exporting country and, if unavailable, a default emission intensity from the EC (2025a) because that is the value that the EU will consider if there is no value.[14] The default emission intensity at the sector level is calculated by aggregating the product-level default emission intensity, weighted by export quantity.[15]

The European Commission data may not accurately reflect the carbon content of MENAAP exporters in CBAM sectors for two reasons. First, estimates represent national average emissions for a product, whereas some exporters will operate with greater or lower carbon efficiency and thus be subject to greater or lower CBAM financial obligations. Note that the EU will derive the CBAM financial obligations based on the emissions of the facility producing that product. If such data are not available, during the early stage of CBAM, exporters might opt for the country-specific default values of emissions—that is, the average emission intensity of the exporting country for that sector from the country-specific default values provided by the EC (2025a).

Second, if country-specific default values are not available for some countries and sectors, we use the global default value provided by the EC (2025a). Because of this, the impacts estimated for those countries and sectors are upper bounds of impacts. For instance, the country-specific default values do not include the data for Egypt's iron and steel sector, so we take the global default emission intensity for that product; however, alternative sources indicate that Egyptian producers outperform the EU average in carbon competitiveness within this sector.[16] MENAAP firms with lower emissions than the default emission intensities will have an incentive to document their actual emissions to avoid the application of the default value and minimize the financial obligations under CBAM. This underscores the importance of developing carbon measurement systems for firms to be able to report accurately to the EU on their emissions.

The CGE analysis offers a stylized simulation of CBAM's impact on changes in trade, welfare, and sectoral output. It captures its impact on exporters through the introduction of CBAM certificate payments and also through changes in prices, especially of fossil fuels, as regional and global actors respond to the removal of free allocations to EU producers and the introduction of CBAM in the EU.

The pre-CBAM policy scenario (the reference scenario against which we measure CBAM impacts) involves the implementation of carbon pricing in the EU at US$100 per tCO_2e alongside output subsidies to EU producers of CBAM goods amounting to 90 percent of the value of carbon content. The value of the subsidy is linked to the sectoral emission intensities, with the biggest impact on fertilizer (45.3 percent) and cement (38.5 percent) for EU producers.

The CBAM policy scenario maintains the US$100 per tCO_2e price and combines the removal of these output subsidies with the introduction of CBAM certificate costs for EU imports. The cost of CBAM certificates is estimated based on the latest default values issued by the European Commission in December (EC 2025a). For several reasons, these simulations are likely to generate impacts at the upper bound:

- They capture the full implementation of CBAM, which in practice will be introduced gradually, with time to allow for improvements in carbon efficiency that are not modeled.
- Exporters are assumed to be using default values, while cleaner exporters will be able to certify their lower emission intensities and face lower CBAM certificate costs.
- The trade reallocation would likely take time, while the model allows for full adjustment within the simulation period.

The AVE payments of CBAM certificate costs range from 2 percent to 236 percent. Carbon payments are calculated as the product of default emission intensity (tCO_2e per ton of product), the assumed carbon price (US$100 per tCO_2e), and export volume to the EU (tons). This approach is consistent with the calculation of the carbon payments in the TEI discussed in the earlier section. The AVE of CBAM cost certificates is calculated as the carbon payment divided by the value of exports to the EU at the sectoral level (refer to table 2.1).[17] The highest AVEs are recorded in cement (which has high emissions intensity and low value per ton of product), and the lowest AVEs are in aluminum (which is liable only for direct emissions and when values per ton of product are high).

Countries facing lower carbon payments will be at a competitive advantage over other exporters. They might also be more competitive than some EU producers depending on the carbon content of EU domestic production. Note that the AVE reported in figure 2.2 differs from that in table 2.1. The distinction arises because the AVE in figure 2.2 is calculated based on relative emission intensity at the product level, whereas the AVE in table 2.1 is aggregated to the sectoral level as defined within the CGE simulation framework. The AVEs from table 2.1 determine the carbon payments incurred by exporters, while EU producers face the price of US$100 per tCO_2e and the removal of free allocations.

TABLE 2.1

Total Carbon Payments as a Share of Export Values to the EU, by Sector, MENAAP Countries

Sector	Economy	Total carbon payment (% of exports to EU)
Aluminum	Egypt, Arab Rep.	9
	United Arab Emirates	7
	Bahrain	7
	Iran, Islamic Rep.	7
	Oman	7
	Saudi Arabia	6
	Qatar	6
	Jordan	2
	Tunisia	2
	Morocco	2
Cement	Saudi Arabia	236
	Algeria	184
	Morocco	137
	Egypt, Arab Rep.	120
	Pakistan	115
	Tunisia	114
	United Arab Emirates	101
Fertilizer	United Arab Emirates	58
	Bahrain	52
	Oman	41
	Algeria	37
	Qatar	36
	Saudi Arabia	31
	Egypt, Arab Rep.	29
	Jordan	17
	Morocco	11
	Tunisia	10

(continued)

TABLE 2.1

Total Carbon Payments as a Share of Export Values to the EU, by Sector, MENAAP Countries (*continued*)

Sector	Economy	Total carbon payment (% of exports to EU)
Iron and steel	Egypt, Arab Rep.	52
	Oman	52
	Saudi Arabia	45
	Algeria	45
	Pakistan	35
	Iran, Islamic Rep.	35
	Tunisia	28
	United Arab Emirates	24
	Kuwait	24
	Morocco	19
	Bahrain	17
	Qatar	15

Sources: Trade data for 2023 using International Trade Analysis Database (BACI) HS2022 version corresponding to Global Trade Analysis Project (GTAP) sector; 2025 default emission intensity data from EC 2025a.
Note: Carbon payments for each country and sector are calculated as follows: default emission intensity (tons per ton) × carbon price (US$100 per ton) × exports to the EU (tons) ÷ exports to the EU (US$) × US$100 per ton. EU = European Union; MENAAP = Middle East, North Africa, Afghanistan, and Pakistan.

CGE MODELING RESULTS

CBAM's implementation is likely to reduce EU imports overall, especially in carbon-intensive sectors, but to increase fertilizer and secondary iron and steel imports. The removal of free allocations in the EU and imposition of financial obligations linked to CBAM lead to

- *Greater embedded carbon costs*, increasing costs of carbon-intensive products from the EU and reducing demand for carbon-intensive imports into the EU;
- *Less demand for fossil fuels (FFLs)*, reducing their prices in the EU and deteriorating the terms of trade of FFLs exporters while benefiting FFLs importers; and
- *Slower growth in the EU* due to higher domestic prices and loss of competitiveness in export markets, reducing demand for imports overall.

With the full implementation of CBAM, EU imports from all countries are anticipated to decrease by US$136 billion, representing a 2.2 percent reduction. Imports specifically within CBAM sectors could drop by US$16 billion, or 8.1 percent. Notably, the EU is expected to see declines in imports of primary aluminum (–US$1.3 billion, –4.5 percent); secondary aluminum (–US$0.3 billion, –1.2 percent); cement (–US$1.1 billion, –34.2 percent); and primary iron and steel (–US$14.4 billion, –16.8 percent). Conversely, imports of fertilizer are expected to rise by US$0.5 billion (6.3 percent), while secondary iron and steel may see an increase by US$0.2 billion (0.5 percent).

CBAM Impacts on MENAAP GDP and Welfare

Modeling results indicate that the impacts of CBAM on GDP and household welfare in MENAAP countries are likely to be negative but negligible. When CBAM is fully implemented, the MENAAP region is projected to experience a net GDP loss of US$1.4 billion, representing a 0.1 percent decline relative to the baseline (a business-as-usual scenario), while the estimated welfare loss is US$6.6 billion (refer to figure 2.3, panel b).[18] Among MENAAP countries, Saudi Arabia and Egypt show large macroeconomic impacts:

- *Saudi Arabia* may face a GDP decrease of US$281 million (0.04 percent loss) and a welfare loss of US$2.2 billion.
- *Egypt* is estimated to experience a GDP decline of US$355 million (0.2 percent loss) and a welfare loss of US$344 million.

Two countries that benefit from increased mitigation efforts in the EU and imposition of CBAM are Jordan and Morocco, though the gains in GDP are very small (0.05 and 0.02 percent, respectively). The gains are driven by the relative carbon competitiveness of their CBAM product exports, as reflected by low AVEs of carbon payments and also because both countries are net importers of FFLs.

CBAM Impacts on MENAAP Exports to the EU

MENAAP's exports to the EU are expected to decline as a result of CBAM. Total MENAAP exports to the EU are projected to decline by US$6.9 billion, representing a 4.2 percent decrease relative to the baseline. The CGE modeling projects that several CBAM-targeted sectors in MENAAP may experience export declines to the

EU: cement (–US$11 million, 60 percent loss); primary iron and steel (–US$26 million, 62 percent decrease); and non-CBAM products (–US$12.5 billion, 8 percent decrease) (refer to figure 2.4). Fertilizer exports to the EU, however, are projected to rise by US$4 billion (231 percent increase) because of the region's relative carbon competitiveness. Non-CBAM exports to the EU drop, but this is partially offset by exports to other markets. Annex 2D provides disaggregated results of MENAAP's countries' non-CBAM exports to the EU and the rest of the world.

FIGURE 2.3

Projected Impact of CBAM on Selected Macro Indicators in MENAAP Countries

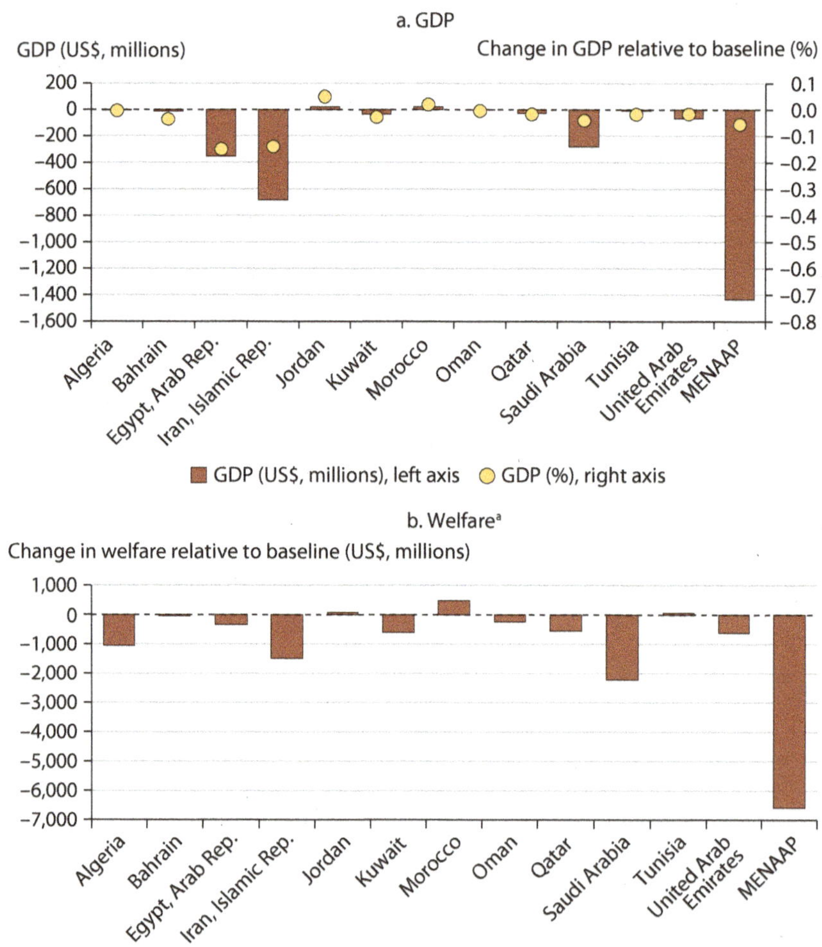

(continued)

FIGURE 2.3

Projected Impact of CBAM on Selected Macro Indicators in MENAAP Countries (*continued*)

c. Terms of trade[b]

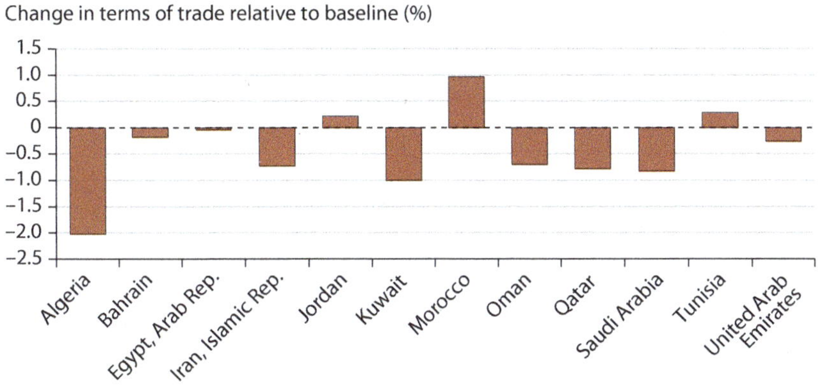

Source: World Bank calculations.

Note: Graphs show the results of computable general equilibrium (CGE) modeling of impacts after the European Union's full Carbon Border Adjustment Mechanism (CBAM) implementation relative to the baseline (business-as-usual) scenario. Afghanistan and Pakistan have been removed from the charts because their change in export value did not exceed the threshold although they are included in overall CGE simulation. GDP = gross domestic product; MENAAP = Middle East, North Africa, Afghanistan, and Pakistan.

a. Welfare measures the amount of income a consumer would require at the original prices to reach the same level of satisfaction they would have after the policy change.

b. Terms of trade refer to the ratio between a country's export price index and its import price index. A relative percentage in change of terms of trade indicates how a policy shock affects a country's welfare by changing the purchasing power of its exports in international markets. area.

Despite an overall decline of exports to the EU, MENAAP exports to the EU will grow in a few CBAM sectors (refer to annex 2E). Increases in primary aluminum exports are recorded by the United Arab Emirates (US$1.2 billion or 12 percent increase), while increases in secondary aluminum exports are driven by Egypt (US$20 million or 11 percent increase) and Saudi Arabia (US$16 million or 15 percent increase). In the fertilizer sector, meanwhile, Morocco (US$2.2 billion or 405 percent increase) and Egypt (US$844 million or 123 percent increase) will see the largest export gains, followed by Tunisia (US$676 million or 398 percent increase). The large fertilizer export gains of Morocco and Egypt are driven by lower prices supported by low emission intensity relative to the EU average and other exporters.

Many countries see very high ad-valorem carbon payments in the cement sector as well as in the iron and steel sector and therefore a decline of exports to the EU under CBAM. The model suggests that Morocco's exports of cement could decline by US$5 million or 76 percent, while Tunisia's exports of cement could fall by US$6.1 million or 55 percent. Even if many other countries see an increase in cement exports to the EU, the overall EU imports of cement from MENAAP decline by 60 percent (refer to figure 2.4). Some of the biggest declines of exports of primary iron and steel are recorded by Algeria and Egypt, with declines of US$4 million (68 percent), and US$14 million (78 percent), respectively (as shown in annex 2E).

FIGURE 2.4

Projected Change in MENAAP's Exports to EU and Rest of the World Relative to Baseline

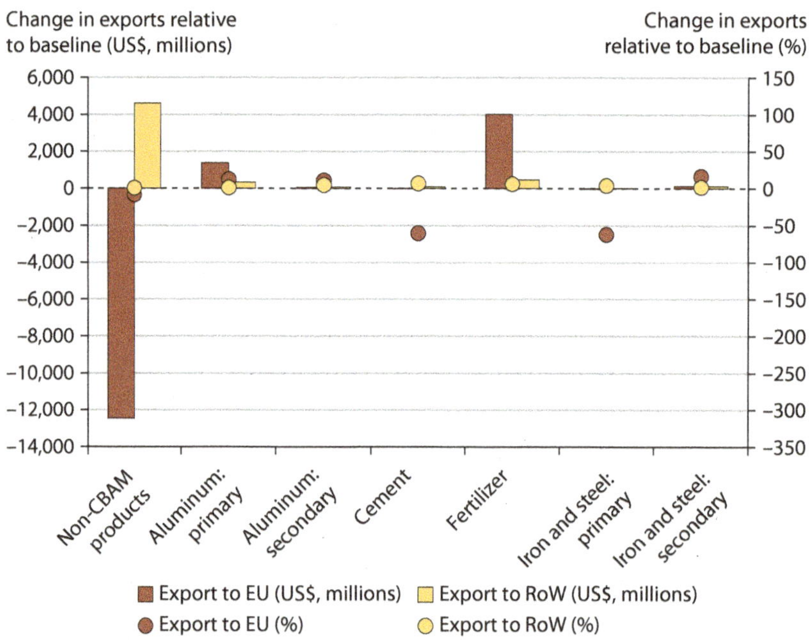

Source: World Bank calculations.
Note: The results are based on the computable general equilibrium (CGE) modeling of the impacts of full Carbon Border Adjustment Mechanism (CBAM) implementation on exports. The baseline is a business-as-usual scenario. The bars show changes in US$, millions and the circles, percentage changes. "Aluminum: primary" refers to primary aluminum, which is new metal extracted directly from raw natural resources. "Aluminum: secondary" refers to recycled metal produced by melting down existing aluminum scrap. "Primary" and "secondary" iron and steel are defined similarly. EU = European Union; MENAAP = Middle East, North Africa, Afghanistan, and Pakistan; RoW = rest of the world.

In the modeling, although total MENAAP exports to the EU could decrease by US$6.9 billion (a 4.2 percent decrease), overall MENAAP exports to the world decline by only US$1.2 billion (a 0.1 percent decrease) as trade rebalances from EU to non-EU markets. The increase in MENAAP exports to other regions is primarily attributed to higher shipments to China, Türkiye, and the United States as well as high-income countries in Asia. Notably, the actual trade redirection may be less significant than model projections suggest because the scenario does not consider the UK's introduction of its own CBAM (as discussed in annexes 2A and 2B), which would disincentivize trade shifting toward the UK.

IMPACTS ON FDI INFLOWS

Although operational only by 2026, the EU CBAM's impact on FDI inflows into the region appears already visible. Data on greenfield FDI since CBAM's announcement in December 2019[19] reveals that Chinese FDI into MENAAP's aluminum and steel sectors has strategically increased while EU investment has declined, suggesting a preemptive response to the upcoming carbon regulations.

The Rise of Chinese FDI

Before the announcement, 43 percent of FDI flows into MENAAP's aluminum sector came from the EU, 26 percent from the rest of the world, and 20 percent from within MENAAP (refer to figure 2.5). After the announcement, China leads at 43 percent, while the EU declined to 21 percent. Meanwhile, Chinese FDI in steel rose from 8 percent to 12 percent during the period, and the EU's share dropped from about 32 percent to 18 percent. Fertilizer and cement saw EU-originated FDI decrease by 13 and 6 percentage points, respectively.

This rising relative importance of China is likely driven less by a surge in new Chinese inflows than by the country's resilience amid shifting global dynamics. Chinese investors appear to be maintaining their long-term positions and exploring MENAAP as a strategic platform for future exports, whereas European investors, facing uncertainty and tightening climate-related policies at home, are becoming more cautious. Indeed, the five-year average of announced EU greenfield projects in CBAM sectors globally has declined by about 16 percent since the Green Deal announcement in 2019. In effect, China's steadier investment trajectory contrasts with the EU's pullback, amplifying China's share in MENAAP's aluminum and steel sectors.

FIGURE 2.5

Chinese Greenfield FDI Flows in MENAAP Have Risen in Aluminum and Steel Sectors Since CBAM's Announcement

Share of FDI inflows, by country of origin (%)

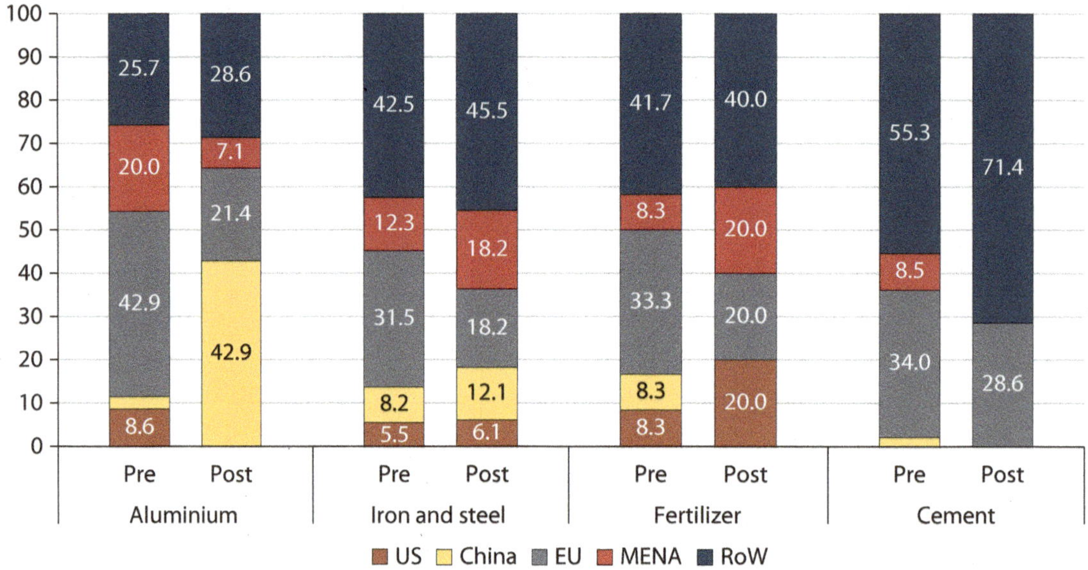

Source: World Bank calculations based on fDi Markets greenfield investment data.
Note: "Pre-CBAM" includes 2010–18 and "Post-CBAM" includes 2019–23. EU = European Union; FDI = foreign direct investment; MENAAP = Middle East, North Africa, Afghanistan, and Pakistan; RoW = rest of the world.

Greenfield FDI Determinants: An Econometric Analysis

The 2019 CBAM announcement under the EU Green Deal seems to have prompted investors to anticipate cost changes on exports and thus on investors' returns—and to adjust their investment decisions accordingly. To test this, we perform an econometric analysis of the determinants of greenfield FDI inflows across countries, sectors, and time, and examine whether investors were more sensitive about a sector's trade exposure to CBAM (as measured by the CBAM TEI) after the CBAM was announced in 2019. For the regression specifications, see box 2.1.[20]

The regression results across all countries indicate that FDI announcements were significantly more sensitive to a sector's CBAM TEI after CBAM was announced than before. A sector's high TEI has a negative impact on FDI announcements, which is to be expected: the greater the trade exposure, the higher the CBAM costs on exports and

BOX 2.1

Estimating the Effect of the CBAM Announcement on FDI

For the econometric analysis of the Carbon Border Adjustment Mechanism (CBAM) impact on greenfield foreign direct investment (FDI), we run the following regression (equation B2.1.1):

$$FDI_{o,d,s,t} = \beta_1 \, Index_{d,s} \times I^{t \geq 2019} + \beta_2 \log(DIST_{o,d}) + \beta_3 \, COMLANG_{o,d} + \beta_4 \log(GDP_o) + \beta_5 \log(GDP_d) + \delta_o + \gamma_d + \phi_t + \varepsilon_{d,s,t}, \quad \text{(B2.1.1)}$$

where $FDI_{o,d,s,t}$ is the long-transformed FDI greenfield investment announcements for sector *s* in year *t*, from country of origin *o* to destination country *d*. FDI is measured both in number of projects and estimated capital expenditure (CAPEX). CAPEX values are deflated by the US Consumer Price Index (CPI) found in the World Development Indicators Database.

$Index_{d,s}$ is the log transformation of CBAM Trade Exposure Index for country *d* in sector *s*. $I^{t \geq 2019}$ is a dummy that turns to 1 in 2019 and onward, signaling the announcement of CBAM within the European Union Green Deal in 2019. The interaction measures the marginal effect of the CBAM exposure index post-2019.

Gravity variables $DIST_{o,d}$ and $COMLANG_{o,d}$ are distance (measured in kilometers) and a dummy variable indicating if countries *o* and *d* share the same language. *GDP* represents the real gross domestic product in constant US dollars. d_o, g_d and ϕ_t are origin, destination, and year fixed effects.

The regression model is employed on bilateral FDI greenfield announcements from the fDi Markets database for the period 2010–23.

the less competitive the sector is and thus less attractive to investment. Globally, the negative effect of a sector's CBAM TEI on FDI is larger and more significant in the aluminum and fertilizer sectors than in the other sectors (refer to figure 2.6). These results suggest that forward-looking investors adjusted their decisions in anticipation of changes in export costs because of CBAM.[21]

In the case of MENAAP, the increased sensitivity of FDI to the CBAM TEI after the CBAM announcement is significant only for aluminum (refer to figure 2.7), and even then, the effect is weaker than in other regions (refer to annex 2F, table 2F.1). In contrast, the marginal impact of the TEI for aluminum on FDI announcements post-2019 is much larger and statistically significant for Latin America and the Caribbean, Sub-Saharan Africa, and South Asia. For fertilizer, the global decline is largely driven by investors in

North America. For cement, the marginal impacts are large and significant for East Asia and the Pacific and South Asia, whereas in MENAAP, the declines are only marginally significant and comparatively small. This relatively muted response of exposed sectors in MENAAP after 2019 may reflect the region's recent sustainability initiatives and enhanced reporting practices (Benny 2024; Kairuz and Goodman 2024).[22]

FIGURE 2.6

Globally, the Negative Effect of a Sector's CBAM TEI on FDI is Largest in the Aluminum and Fertilizer Sectors After the 2019 CBAM Announcement Within the EU Green Deal

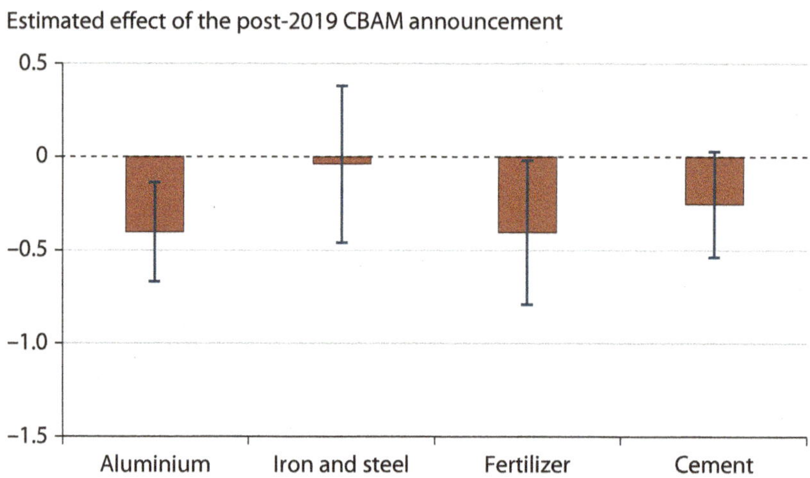

Source: World Bank calculations based on fDi Markets data.
Note: Each column shows results from a gravity model regression of foreign direct investment (FDI) greenfield investment announcements on the Trade Exposure Index (TEI) interacted with post-2019 dummy on log capital expenditure (in US$, millions) by country pair (greenfield FDI from country A in country B) and year, with year, source, and destination fixed effects (as further detailed in box 2.1). This figure presents pooled results across all countries, whereas figure 2.7 includes region-specific interactions, highlighting the estimates for the Middle East, North Africa, Afghanistan, and Pakistan (MENAAP). Whiskers indicate 95 percent confidence intervals. A 0.25 estimate implies (e^(0.25) – 1)*100 = 28 percent higher announced greenfield FDI in response to a percent increase in the TEI relative to its pre-2019 level. MENAAP countries included in the sample are Algeria; Bahrain; Egypt, Arab Rep.; Iran, Islamic Rep.; Iraq; Jordan; Kuwait; Lebanon; Morocco; Oman; Qatar; Saudi Arabia; Syrian Arab Republic; Tunisia; and West Bank and Gaza. CBAM = Carbon Border Adjustment Mechanism; EU = European Union.

FIGURE 2.7

Globally, The Negative Effect of a Sector's CBAM TEI on FDI Is Not Driven by MENAAP, Except for the Aluminum

Estimated effect of the post-2019 CBAM announcement (%)

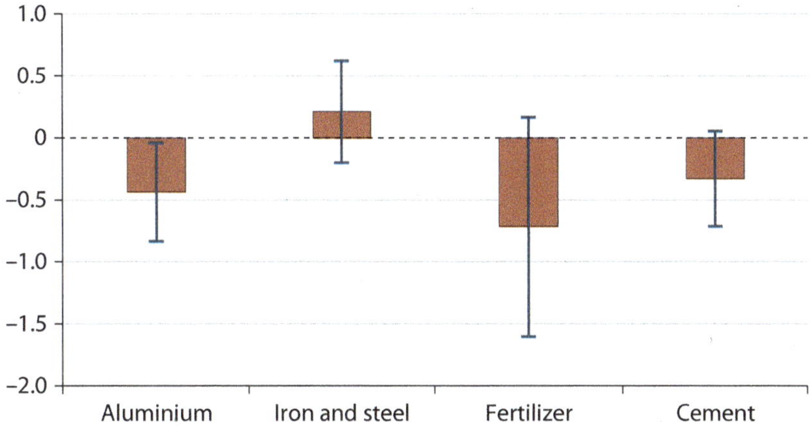

Source: World Bank calculations based on fDi Markets data.
Note: Each column shows results from a gravity model regression of foreign direct investment (FDI) greenfield investment announcements on the Trade Exposure Index (TEI) interacted with post-2019 dummy on log capital expenditure (in US$, millions) by country pair (greenfield FDI from country A in country B) and year, with year, source, and destination fixed effects (as further detailed in box 2.1). Whereas figure 2.6 presents pooled results across all countries, this figure includes region-specific interactions, highlighting the estimates for the Middle East, North Africa, Afghanistan, and Pakistan (MENAAP). Whiskers indicate 95 percent confidence intervals. A 0.25 estimate implies (e^(0.25) − 1)*100 = 28 percent higher announced greenfield FDI in response to a percent increase in the TEI relative to its pre-2019 level. MENAAP countries included in the sample are Algeria; Bahrain; Egypt, Arab Rep.; Iran, Islamic Rep.; Iraq; Jordan; Kuwait; Lebanon; Morocco; Oman; Qatar; Saudi Arabia; Syrian Arab Republic; Tunisia; and West Bank and Gaza. CBAM = Carbon Border Adjustment Mechanism; EU = European Union.

TRADE AND FDI VULNERABILITY, BY COUNTRY

MENAAP countries' vulnerability to CBAM is determined by the CBAM TEI and by the FDI exposure. A country's TEI in a sector captures a sector's potential loss or gain of competitiveness in the EU market. A country's FDI exposure captures the presence of FDI in the country and thus of available foreign capital to realize potential competitiveness gains in a sector resulting from CBAM. We measure FDI exposure by the importance of greenfield FDI in a country's domestic capital stock during the pre-CBAM announcement period, 2016–18.

A Framework for Categorizing Trade Vulnerability and FDI Exposure

Countries can be categorized into four quadrants for a given sector based on their trade and FDI exposure (refer to figure 2.8):

- High vulnerability with high trade and FDI exposure (upper-right quadrant)
- High trade but low FDI exposure (upper-left quadrant)
- Low trade but high FDI exposure (lower-right quadrant)
- Low exposure on both fronts (lower-left quadrant).

This framework helps identify countries facing adjustment pressures or opportunities from CBAM. Economies with lower carbon emissions in a sector than the EU average (denoted by a negative CBAM TEI) and high FDI exposure could potentially attract green investments and realize the sector's potential post-CBAM implementation. Economies with lower carbon emissions than the EU average but low FDI exposure will not be able to tap into FDI in the short term to realize the sector's potential and will be constrained to available domestic capital. Economies with carbon emissions higher than the EU but high FDI exposure may leverage FDI to help decarbonize these sectors.

FDI Implications for MENAAP Countries

In MENAAP, most countries with high trade exposure in aluminum and in iron and steel rely only minimally on FDI (refer to figure 2.8), implying that short-term sectoral growth in these areas will need to be driven primarily by domestic capital investment rather than foreign inflows. Notable exceptions stand out: for aluminum, Bahrain, Egypt, and Morocco combine both higher-than-median FDI exposure and a high CBAM trade exposure, whereas most other MENAAP countries fall into the category of high trade but low FDI exposure. For iron and steel, Algeria and Egypt also show both high CBAM trade exposure and high FDI exposure, even though most countries in the region tend to exhibit higher-than-median trade exposure alongside high exposure in FDI. These exceptions (Bahrain, Egypt, and Morocco in aluminum; and Algeria and Egypt in iron and steel) are particularly relevant because their receipt of FDI makes them more directly affected by CBAM-related investment dynamics.

FIGURE 2.8

Most Countries in MENAAP with High Trade Exposure in Aluminum and in Iron and Steel Rely Only Minimally on FDI

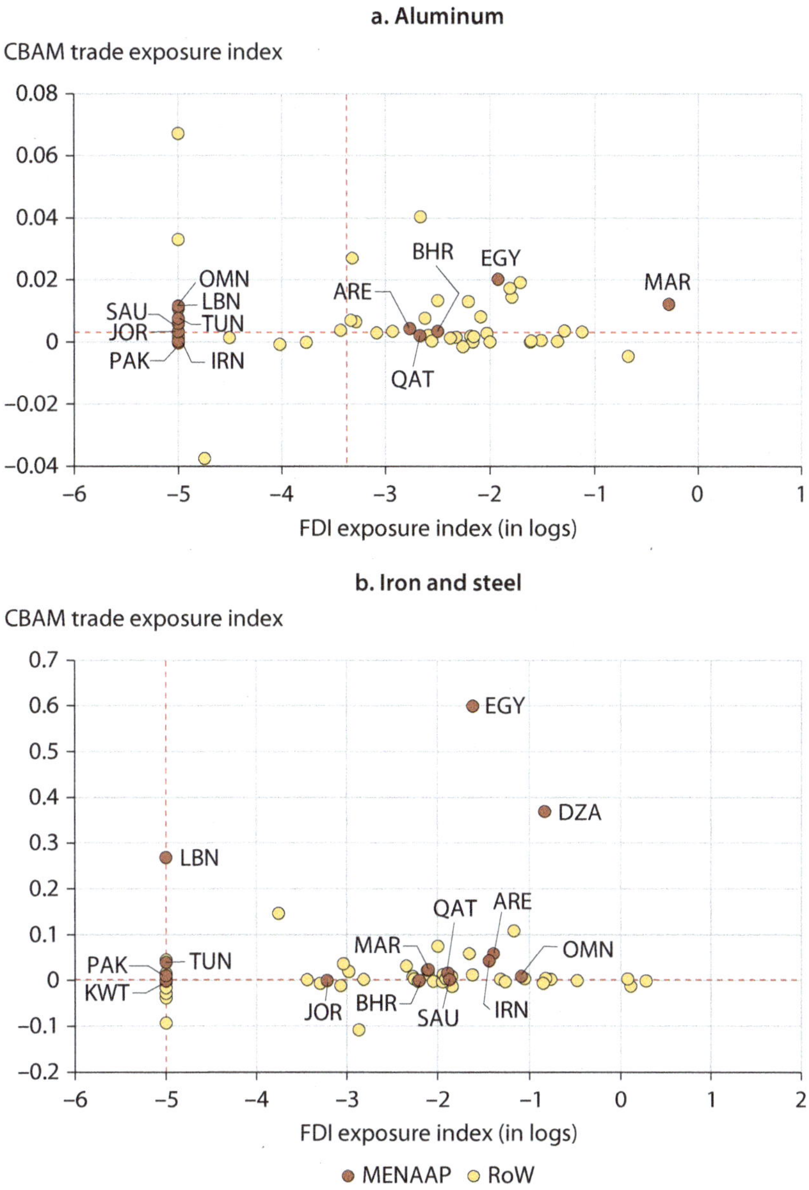

Sources: World Bank calculations based on fDi Markets data and the International Trade and Production Database for Simulation (ITPD-S).

Note: Graphs depict the relation between the Carbon Border Adjustment Mechanism (CBAM) Trade Exposure Index and foreign direct investment (FDI) exposure (in logarithms) measured as the amount of announced capital expenditure in greenfield investment per unit dollar of domestic output between 2016 and 2018. Log transformations are in base 10 and add a constant of 1e-5. Dotted lines represent median values for each variable. For a list of country codes go to https://www.iso.org/obp/ui/#search. MENAAP = Middle East, North Africa, Afghanistan, and Pakistan; RoW = Rest of the World.

A similar situation is observed in the cement and fertilizer sectors. Most countries in these sectors have very limited FDI and low export exposure. However, several countries stand out (refer to figure 2.9). For cement, Algeria, Egypt, Morocco, and Tunisia all exhibit high CBAM trade exposure, but only Algeria and Morocco show high FDI exposure. For fertilizer, Algeria is an outlier with both high CBAM trade exposure and high FDI exposure. Morocco and Oman also have higher-than-median FDI exposure; Oman combines this with higher trade exposure, whereas Morocco's trade exposure is lower than the median.

Across these sectors, some countries are not necessarily FDI-exposed but are trade-exposed (such as Bahrain, Egypt), whereas others (including Jordan, Qatar, Saudi Arabia, and Tunisia) have trade exposure levels below the median. In many other MENAAP countries, FDI accounts for a meaningful share of gross output, yet their export exposure remains minimal. These patterns underscore that only a few countries (such as Algeria and Morocco in cement, and Algeria and Oman in fertilizer) present the combination of trade and FDI exposure that could position them to leverage FDI in adapting to CBAM-related challenges.

FIGURE 2.9

In MENAAP, Most Countries in the Cement and Fertilizer Sectors Have Limited FDI and Low Export Exposure, Though a Few Stand Out

a. Cement

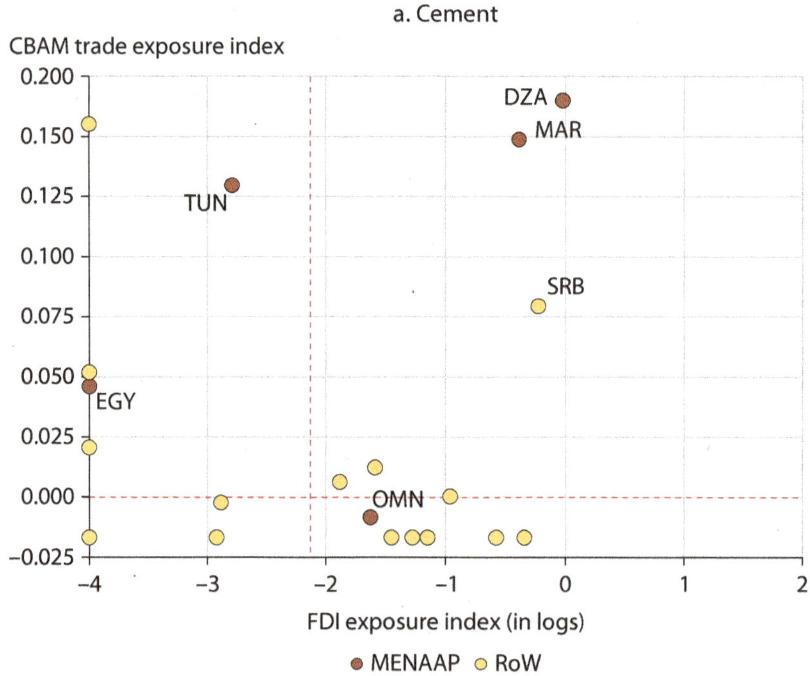

(continued)

FIGURE 2.9

In MENAAP, Most Countries in the Cement and Fertilizer Sectors Have Limited FDI and Low Export Exposure, Though a Few Stand Out (*continued*)

b. Fertilizer

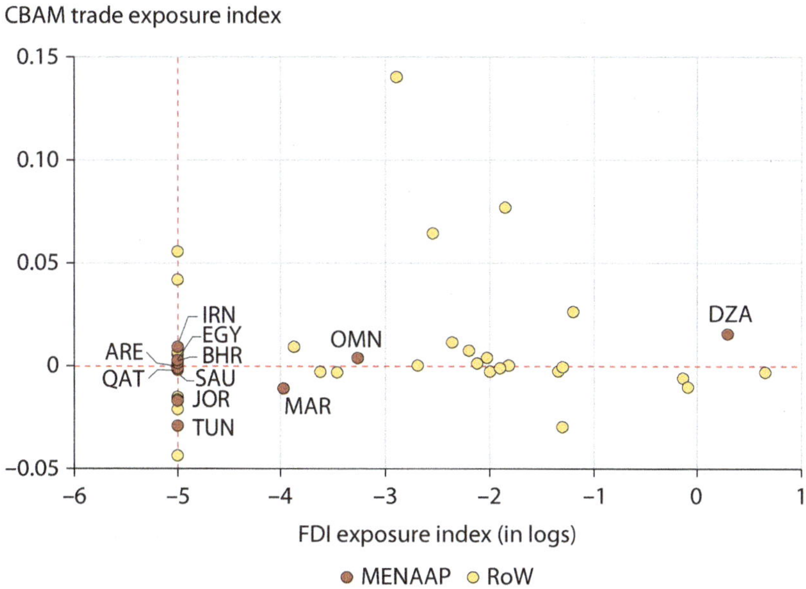

Sources: World Bank calculations based on fDi Markets data and the International Trade and Production Database for Simulation (ITPD-S).
Note: Graphs depict the relation between the Carbon Border Adjustment Mechanism (CBAM) Trade Exposure Index and foreign direct investment (FDI) exposure (in logarithms) measured as the amount of announced capital expenditure in greenfield investment per unit dollar of domestic output between 2016 and 2018. Log transformations are in base 10 and add a constant of 1e-5. Dotted lines represent median values for each variable. International Organization for Standardization (ISO) country codes available at https://www.iso.org/obp/ui/#search. MENAAP = Middle East, North Africa, Afghanistan, and Pakistan; RoW = Rest of the World.

MENAAP firms integrated into EU value chains in sectors such as textiles, agriculture, and raw materials are also subject to the EU CS3D which extends due diligence requirements across participating firms in those chains. This increases sustainability and human rights compliance obligations and may reduce EU market access for affected MENAAP firms (see box 2.2).

BOX 2.2

EU Corporate Sustainability Due Diligence

The Corporate Sustainability Due Diligence Directive (CS3D) of the European Union (EU), effective July 2024, requires large companies with over 1,000 employees and annual net turnover above €450 million (globally for EU firms or within the EU for foreign firms) to address human rights and environmental impacts across their operations and value chains in sectors such as textiles, agriculture, and raw materials. Companies must implement due diligence processes to prevent, mitigate, and, where necessary, remediate negative impacts on human rights and the environment, with penalties for noncompliance. Additionally, they must develop plans to align with the 1.5 degrees Celsius global warming limit, setting climate targets for 2030 and every five years up to 2050 based on scientific evidence.

The regulation will have the greatest impact on countries whose companies generate significant export revenues in the EU or are integrated into EU value chains and do not comply with human rights and environmentally related requirements. The global value chain (GVC) exposure data reveal that only a small share of exporting firms in the Middle East, North Africa, Afghanistan, and Pakistan (MENAAP) region will be directly subject to the directive, but a large share will be affected indirectly (that is, exports to non-EU countries that are ultimately destined to the EU through supply chains). (For the agriculture and mining and quarrying sectors, refer to figure B2.2.1 and for the remaining sectors, refer to annex 2G).

Libya's GVC exposure to the EU CS3D records 64 percent of gross mining and quarrying output and 21 percent of gross metal products output. Morocco is the most exposed in agriculture and Tunisia in petroleum, chemical and nonmetallic minerals, and textile and apparel. This trade exposure reflects MENAAP countries' comparative advantages, their trade relationships with the EU (for direct exports), and their integration into global supply chains (for indirect exports). Estimating the impact of the EU CS3D on a specific country and sector would require a detailed firm analysis capturing firms' gaps relative to EU requirements, adjustment costs, and their implication on sales.

The CS3D's rigorous environmental and human rights due diligence requirements might initially slow down EU investments in MENAAP. However, it could lead to more responsible foreign direct investment (FDI) as companies upgrade their supply chains and practices to meet EU standards. This may boost FDI in sectors meeting EU sustainability criteria and reduce investments in noncompliant industries.

(continued)

BOX 2.2 (*continued*)

FIGURE B2.2.1

Mining and Quarrying Is the MENAAP Region's Most Exposed Sector to EU CS3D

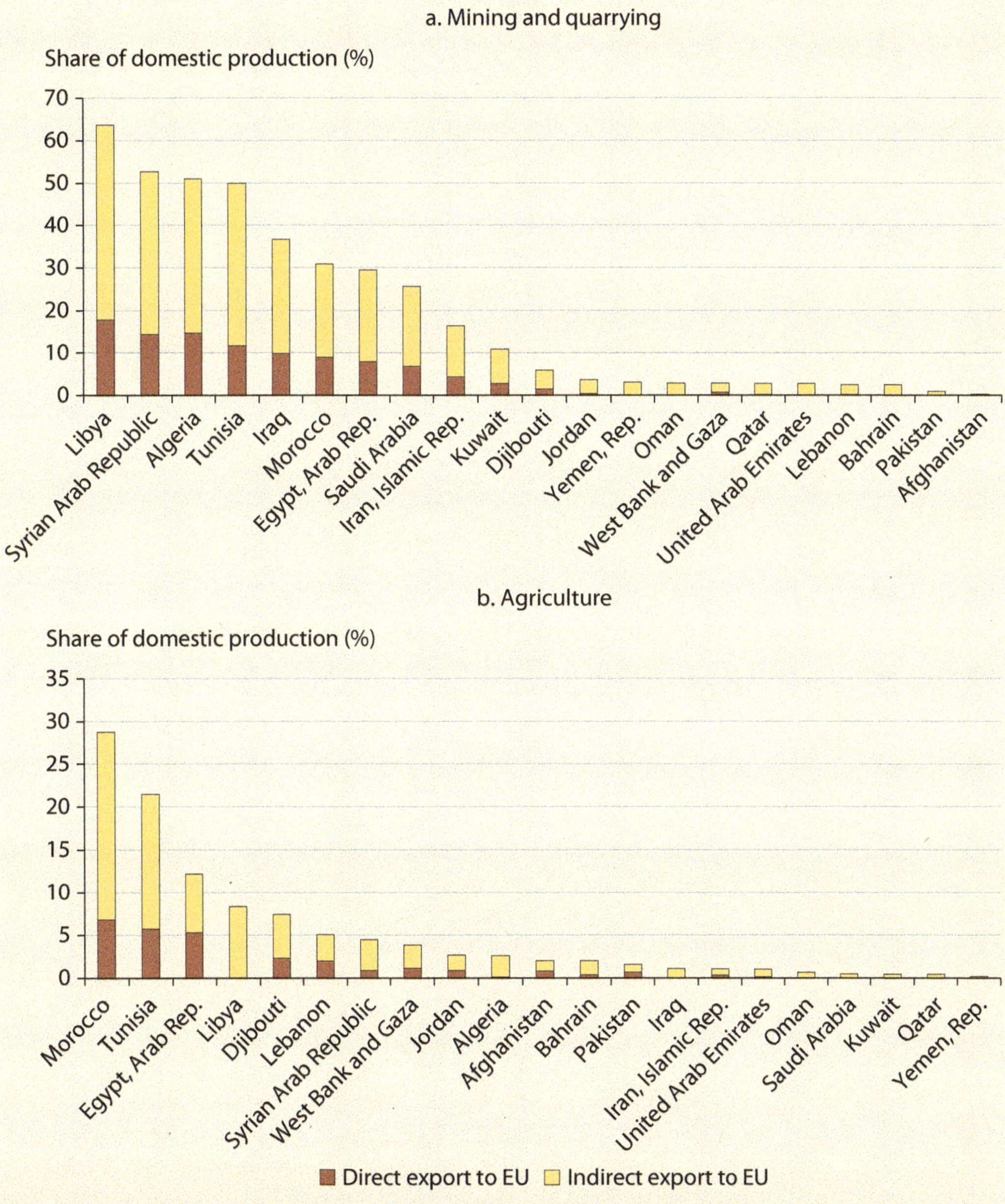

Sources: World Bank calculations and Arvis et al. 2025.

Note: The global value chain (GVC) exposure index follows the indicator developed by Arvis et al. (2025) using the methodology of Baldwin et al. (2022). Using the 2022 EORA global supply chain database, GVC exposure to the European Union (EU) is the sum of direct exports to the EU and indirect exports to the EU over domestic output of products. CS3D = Corporate Sustainability Due Diligence Directive; MENAAP = Middle East, North Africa, Afghanistan, and Pakistan.

NOTES

1. This analysis covers a sample of 14 countries in the MENAAP region: Algeria; Bahrain; Egypt, Arab Rep.; the Islamic Republic of Iran; Jordan; Lebanon; Libya; Morocco; Oman; Pakistan; Qatar; Saudi Arabia; Tunisia; and the United Arab Emirates.
2. "Direct emissions" (scope 1) refers to emissions generated directly from the production processes of CBAM goods. They include emissions from the production of heating and cooling used during these processes, regardless of whether the heating or cooling is produced on-site or off-site. "Indirect emissions" (scope 2) are the indirect emissions that come from purchased energy such as electricity, heat, or cooling by the company.
3. Note that credits are provided for direct taxes on carbon, not for indirect carbon taxation. Fuel excise taxes are an indirect carbon taxation because the tax is on the value, not on the carbon content.
4. From 2026 to 2034, there will be a gradual phaseout of ETS free allowances for CBAM sectors. CBAM will be gradually phased in according to the following factors: 0 percent in 2026, 2.5 percent in 2027, 5 percent in 2028, 22.5 percent in 2029, 48.5 percent in 2030, 61 percent in 2031, 73.5 percent in 2032, and 85 percent in 2033. From 2034, free allowances will be fully phased out, and the CBAM will apply at 100 percent, meaning importers will bear the full carbon cost as if they were under the ETS.
5. The European Commission provides a markup that is an extra percentage added to the default value so that using defaults is deliberately more conservative (higher) than using verified actual emissions, which encourages reporting installation-specific data.
6. As in the case of a firm, a sector's exposure to CBAM is determined by its relative emission intensity and its export orientation to the EU.
7. To prevent any misinterpretation, the analysis includes countries whose export values exceed US$1 million for each CBAM sector.
8. BACI is a database of the Centre for Prospective Studies and International Information (CEPII): https://www.cepii.fr/CEPII/fr/bdd_modele/bdd_modele_item.asp?id=37.
9. The analysis was done by January 2026.
10. Emission intensities can fluctuate over time, vary across firms, and depend on data collection methods and sample coverage, which may affect the accuracy and interpretation of competitiveness outcomes.
11. The sample used in the CGE modeling includes 12 countries: Algeria, Bahrain, Egypt, the Islamic Republic of Iran, Jordan, Kuwait, Morocco, Oman, Qatar, Saudi Arabia, Tunisia, and the United Arab Emirates.
12. Emission intensity data are from European Commission (2025a), and the model is from McDougall and Golub (2007).
13. The EU grants free allowances to its domestic producers, so they pay the ETS carbon price only on production exceeding those allowances, and these free allocations are to be phased out as CBAM is implemented.

14. Because raw emission intensity is provided at Harmonized System (HS) 6-digit level by European Commission (2025a), the emission intensity of sectors such as aluminum, cement, fertilizer, and iron and steel was weighted by export quantity to the EU.
15. In the case of sectors for which emission data are available but are outliers, values are replaced with trade weighted regional average emission intensity.
16. The alternative data are from an independent monitor of GHG emissions data, Climate TRACE (Tracking Real-Time Atmospheric Carbon Emissions), https://climatetrace.org/.
17. The concepts of total carbon payment in table 2.1 and in figure 2.2 (panels a, d, g, and j) are the same, but the numbers vary slightly because the CGE analysis only covers raw and semifinished or finished products, whereas the data in figure 2.2 cover all HS codes from the CBAM legislation, including downstream products such as nuts and bolts.
18. Welfare or equivalent variation is a monetary measure of how much a consumer would be willing to pay to avoid a price change, or conversely, how much they would need to be compensated for a price change to maintain the same level of utility. It's essentially the change in income that would produce the same welfare effect as the price change.
19. In December 2019, the European Commission proposed a CBAM for selected sectors as part of the European Green Deal (European Parliament Legislative Train, 2025). For additional details, go to https://www.europarl.europa.eu/legislative-train/theme-a-european-green-deal/file-carbon-border-adjustment-mechanism.
20. The regression analysis focuses on potential anticipation effects of CBAM on FDI announcements, using the 2019 discontinuity when the program was first announced. The analysis does not account for subsequent design changes to the EU ETS and CBAM between 2019 and 2025, nor for fluctuations in the EU ETS price, which determine the effective cost of CBAM for firms. Future work could build on these dimensions by exploiting variations in ETS price movements or policy reforms as exogenous shocks, as well as by controlling for sectoral differences in energy efficiency and broader shocks such as the COVID-19 pandemic, to strengthen identification.
21. Meng (2017) uses a similar approach, leveraging the failed Waxman-Markey Bill in the United States during 2009–10 to assess stock returns on firms whose carbon emissions are just around the threshold. Kruse et al. (2024) use the signing of the Paris Agreement as a quasi-natural experiment to evaluate the stock market response to firms engaged in green activities. Both studies observed significant effects on stock returns. This suggests that policy announcements can influence markets before implementation and that market participants may possess local information, which can provide sufficient statistics to evaluate the potential benefits or costs of the policy.
22. The Gulf States' pivot to low-carbon metals manufacturing positions them advantageously ahead of EU carbon tariffs in 2026 (Benny 2024).

REFERENCES

Arvis, J.-F., A. Burman, A. Espitia, J.-C. Maur, N. Rocha, and D. Ulybina. 2025. "Measuring Exposure and Vulnerability to International Trade Shocks." Policy Research Working Paper 11243, World Bank.

Baldwin, R., R. Freeman, and A. Theodorakopoulos. 2022. "Horses for Courses: Measuring Foreign Supply Chain Exposure." Working Paper 30525, National Bureau of Economic Research.

Benny, J. 2024. "Gulf States in Race to Make Low-Carbon Metals as EU Carbon Tax Looms." *The National*, October 5. https://www.thenationalnews.com/business/economy/2024/10/05/gulf-states-in-race-to-make-low-carbon-metals-as-eu-carbon-tax-looms/.

Burnett, N., A. Dillon, T. Hewitt, and I. Jozepa. 2026. "Carbon Border Adjustment Mechanism." Research Briefing 9935, House of Commons Library, UK Parliament.

DESNZ (Department for Energy Security and Net Zero). 2023. "Factsheet: UK Carbon Border Adjustment Mechanism." Consultation outcome, DESNZ, UK Treasury.

EC (European Commission). 2025a. "Annexes to the Commission Implementing Regulation (EU) 2025/2621 of 16 December 2025 Laying Down Rules for the Application of Regulation (EU) 2023/956 of the European Parliament and the Council as Regards the Establishment of Default Values." European Commission.

EC (European Commission). 2025b. "Proposal for a Regulation of the European Parliament and of the Council Amending Regulation (EU) 2023/956 as Regards Simplifying and Strengthening the Carbon Border Adjustment Mechanism." COM(2025)87, February 26, European Commission.

EU (European Union). 2025. "Regulation (EU) 2025/2083 of the European Parliament and of the Council of 8 October 2025 Amending Regulation (EU) 2023/956 as Regards Simplifying and Strengthening the Carbon Border Adjustment Mechanism." *Official Journal of the European Union*, October 17, 2025.

ICAP (International Carbon Action Partnership). 2024. "UK Outlines Details for Carbon Border Adjustment Mechanism Introduction from 2027." ETS News, November 13. https://icapcarbonaction.com/en/news/uk-outlines-details-carbon-border-adjustment-mechanism-introduction-2027.

Kairuz, M., and K. Goodman. 2024. "Uniting for Sustainability: How Regional Cooperation Can Keep MENA Businesses Competitive." *Forum Stories*, October 21, World Economic Forum.

Kruse, T., M. Mohnen, and M. Sato. 2024. "Do Financial Markets Respond to Green Opportunities?" *Journal of the Association of Environmental and Resource Economists* 11 (3): 549–76. doi:10.1086/727370.

Maliszewska, M., C. Fischer, E. Jung, and M. Chepeliev. 2025. "Carbon Border Adjustment Mechanism (CBAM) Exposure Indices Methodological Note." Prosperity Note 202614, World Bank.

McDougall, R., and A. Golub. 2007. "GTAP-E: A Revised Energy-Environmental Version of the GTAP Model." GTAP Research Memorandum 15, Global Trade Analysis Project.

Meng, K. C. 2017. "Using a Free Permit Rule to Forecast the Marginal Abatement Cost of Proposed Climate Policy." *American Economic Review* 107 (3): 748–84. doi:10.1257/aer.20150781.

Vidovic, D., A. Marmier, L. Zore, and J. A. Moya. 2023. *Greenhouse Gas Emission Intensities of the Steel, Fertilisers, Aluminium and Cement Industries in the EU and Its Main Trading Partners.* Joint Research Centre Technical Report, Publications Office of the European Union.

Yotov, Y. V., R. Piermartini, J.-A. Monteiro, and M. Larch. 2016. "An Advanced Guide to Trade Policy Analysis: The Structural Gravity Model." United Nations and World Trade Organization.

CHAPTER 3

Building Resilience and Seizing Economic Opportunities

INTRODUCTION

The Middle East, North Africa, Afghanistan, and Pakistan (MENAAP) region is highly exposed to climate-related issues. Stressors have been affecting the region for a long time, including the following:

- *Freshwater availability* is now below the "absolute water scarcity" threshold in more than half of the countries in the region.[1]
- *Rising temperatures* intensify desertification in the Arab Republic of Egypt, Iraq, Jordan, Pakistan, and the Republic of Yemen, contributing to crop losses and higher energy demand for cooling. Extreme heat (days whose maximum temperature exceeds 35 degrees Celsius) adversely affects firms, leading to decreased productivity, reduced jobs, and increased operational costs (as discussed in chapter 1).
- *Longer and more severe droughts* threaten the rainfed agriculture that supports rural livelihoods, also leading to reductions in sales, labor productivity, and investment at the firm level (Gatti, Onder et al. 2025).

- *Flooding and sea level rise* pose growing risks: For example, Egypt's Nile Delta alone could see damages equivalent to 1 percent of gross domestic product (GDP) annually by 2030 due to sea level rise (World Bank 2022a).

These hazards, documented across World Bank Country Climate and Development Reports (CCDRs),[2] will impose mounting economic and social costs. By 2040, under current policies used as a baseline, countries could face potential GDP losses ranging from about 1–2 percent (Lebanon) to 4–5.5 percent (Pakistan), with lower-income economies and those in hotter climates experiencing the greatest impacts. These losses come at a time when exogenous regulatory changes are implemented, such as new trade policies implemented by the European Union (EU): the Carbon Border Adjustment Mechanism (CBAM) and the Corporate Sustainability Due Diligence Directive (CS3D), as discussed in chapter 2.

While complex, such an environment will still provide opportunities to invest and accelerate development in MENAAP. All risks cannot be offset, but high returns can be obtained provided key policy changes are implemented. For example, research suggests that investments in adaptation may create or protect the equivalent of 25 million jobs across a sample of 49 low- and middle-income countries (World Bank 2025d). With this background in mind, this final chapter delves into the policies required to help firms become less vulnerable to increasing temperatures, increase productivity, and comply with new regulatory requirements in the markets of high-income economies, therefore keeping MENAAP economies moving forward.

Three broad types of policies can help in this regard:

- *Improving access to finance (A2F) and the business environment*, especially for smaller firms, will be key. This will require specific new interventions but also continuing to implement foundational policies that deal with more traditional issues in this region (World Bank 2022g, 2025a, 2025c).

- *Supporting firms' "green competitiveness"* by fostering decarbonization will also be needed to ensure continuous market access.

- *Fostering the economic opportunities* that will arise from the move towards increased sustainability will better position the region's firms to seize those opportunities if the suggested policy changes have been implemented.

FINANCING THE ADOPTION OF GREENER TECHNOLOGIES

Meeting the challenge of a changing climate is largely an issue of technological adoption. Measures such as adapting to increased temperatures with improved cooling systems, fostering energy efficiency by using better equipment and processes, having heat-resistant buildings, or creating a "heat-proof" value chain rely on the adoption of new, greener technologies. Similarly lowering carbon emissions from production processes also relies on the use of newer technologies. The adoption of technologies by firms is a complex process that relies on many factors such as the expected returns of a new technology, the readiness of the firm (in terms of human resources and managerial capacities), the intensity of competition, the regulatory landscape, information on available technologies, and the availability of financing (Cirera et al. 2022).

Challenges from Weak Access to Finance

Financing is critical for adopting greener technologies, but the region faces serious A2F challenges. World Bank Enterprise Survey data show that close to 25 percent of firms in the region identify A2F as a "major" or "very severe" constraint, against about 23 percent worldwide. The constraint is more important for smaller firms (refer to annex 3A, figure 3A.1). Small and medium enterprises (SMEs) are especially affected because they receive only 7 percent of banks' credit, with the largest corporates absorbing most of the financing (Dornel et al. 2020).[3]

Weak A2F amplifies the impact of climate shocks, reflecting failures on both the supply and demand sides of financial markets. On the supply side, lending institutions (and increasingly insurance companies) remain reluctant to extend credit (insurance) to firms, particularly the micro, small, and medium enterprises (MSMEs) that constitute the backbone of economies in the region. This reluctance reflects structural weaknesses in credit infrastructure, including weak insolvency regimes, underdeveloped collateral frameworks, and limited credit information systems, combined with antiquated lending practices that continue to rely heavily on collateral-based lending (refer to annex 3A, figure 3A.2). Risk perceptions toward MSMEs are already high, but banks' growing focus on climate-related exposures is further accentuating this bias: firms in locations vulnerable to climate events are increasingly subject to higher interest rates or outright credit rationing.[4]

On the demand side, long-standing bottlenecks such as weak governance, limited transparency in financial reporting, and low levels of equity constrain firms' ability to access financing. These constraints are compounded by limited awareness of adaptation and mitigation options and as mentioned above, by the high up-front costs of available solutions. As a result, climate vulnerability not only raises the cost of capital but also restricts the financing available for investment in resilience (Kling et al. 2021). Addressing these challenges will require scaling up "green" A2F through policy measures aiming at deepening financial intermediation in general and promoting the right set of standards and innovative financial instruments.

Financial Products to Improve A2F

Developing innovative and customized financial instruments will be important to overcome A2F limitations. Financial products for green and low-carbon technologies answer several needs:

- *They can help correct market failures such as environmental externalities.* Because emitters do not bear the full social cost of greenhouse gases (GHGs), private rates of return on such projects are typically too low.
- *They help correct underinvestment in innovation.* Knowledge spillovers often mean firms underinvest in research and development and the adoption of new technologies (Jaffe et al. 2004).
- *They can steer technical change,* countering path dependency by helping economies to move out of established high-carbon technologies (Acemoglu et al. 2012; Aghion et al. 2016).
- *They can help address high up-front costs and long payback times,* because many clean technologies (for renewables, energy efficiency, and industrial decarbonization) are capital intensive and face cash-flow risks related to the technology and the policy environment (IRENA 2016; World Bank 2025b).
- *They help address information asymmetries* as investors often have difficulties in assessing the environmental integrity and financial performance of "green" projects (OECD 2022).

Although green finance largely remains at an early stage in MENAAP,[5] several emerging initiatives suggest that the region has the potential to unlock capital at scale.

Green taxonomies (GTs). These taxonomies are a critical classification tool providing financial markets with information on sustainable business activities and guidelines for identifying them. They can support climate-informed risk analysis and facilitate the certification of green assets. Globally, only about 10 percent of emerging markets and developing economies have GTs in place against 76 percent of high-income economies (World Bank 2024a). By December 2025, only Pakistan had a functioning GT in MENAAP. However, countries such as Egypt, Iraq, Jordan, Morocco, Tunisia, and the United Arab Emirates are currently preparing their own. Besides the technical design of GTs, a broad inclusion of all stakeholders, especially smaller enterprises, is key for adoption, which itself could be made easier if GTs mirror the country's existing economic classifications.

Green and sustainable finance instruments. The issuance of green loans, green bonds (refer to annex 3A, figure 3A.3), sustainability-linked bonds, and transition bonds (see box 3.1), customized to firms' specific characteristics, is important to support firm performance, reduce the cost of capital, and increase transparency (Flammer 2021; Tang and Zhang 2020; Wang et al. 2020; Zerbib 2019). Their use in the region is still limited, mainly driven by state-owned enterprises (SOEs) and out of reach for most SMEs. Still, green sukuk (a type of Islamic bond where the proceeds are used to fund specific environmentally sustainable projects) worth a total of US$6.5 billion were issued in the region in 2023, representing more than half of all global green sukuk (Emirates News Agency 2024). The same year, Saudi Arabia and the United Arab Emirates dominated the green bond league tables, accounting for a combined 77 percent of total issuances in the region.[6]

To ensure greater use, GTs must first be in place, followed by measures to stimulate the markets (such as fee reductions and tax incentives to encourage issuance to SMEs); to ensure market integrity (through appropriate disclosure, reporting, and independent verification and certification mechanisms); and to build capacity while preventing greenwashing through appropriate enforcement measures.

Climate insurance. Climate-driven acute risks (extreme heat events, floods, and droughts) are a major source of uncertainty in the region. They can have consequences spanning capital stock destruction (including building damage), indirect economic losses (such as business interruption), and transmission to the financial sector (for example, nonperforming loans and impact on banks' balance sheets).

BOX 3.1

Different Uses of Green and Sustainability-Linked Bonds

Green bonds have been increasingly issued by firms with clear green projects and are subscribed to by investors who want to align their investments with the goals of the Paris Agreement. For firms that operate in brown sectors and do not want to tie the use of proceeds to specific green projects—which can be the case of many large firms in the Middle East, North Africa, Afghanistan, and Pakistan (MENAAP)—the use of green bonds to decarbonize their value chains can be more constraining because of the reporting requirements. For those firms, the use of sustainability-linked bonds and transition bonds can be more useful.

Sustainability-linked bonds are bonds whose coupon rates are contingent on the achievement of predefined specific sustainability (environmental, social, and governance [ESG]) goals. The coupon rate decreases if the issuer achieves the predefined goals and increases otherwise.

Transition bonds, on the other hand, are issued by firms in carbon-intensive sectors (energy, mining, industry, transportation) to finance their shift toward cleaner activities without necessarily linking the coupon to specific sustainability targets. Those companies are usually in the early stage of their low-carbon transition and lack mature green projects that can be eligible for green bond issuance.

To further develop these markets in MENAAP, it is essential to encourage the issuance of local currency-labeled bonds, which can help diversify the investor base and reduce the currency risk and the cost of capital while supporting small and medium enterprises' access to green finance.

Developing climate and disaster risk financing instruments and frameworks—in particular, insurance—can help reduce the subsequent financial burden on firms. It can also help de-risk firms' adaptation investments. Banks may be more likely to lend to firms with climate insurance because they will be protected for potential losses. The provision of these instruments should be part of broader financial packages, combining them with savings, payments, and credit instruments within a consolidated "A2F" approach.

Yet, supply alone is inefficient if not met with demand, and national regulators should work with the insurance sector to sensitize and build firms' capabilities to better understand how insurance can help de-risk their assets and protect their revenues from climate risks. Overall, close collaboration in the form of public-private insurance programs can help operationalize climate insurance programs (Adam et al. 2025).

Concessional finance. Concessional finance (capital offered at more favorable terms than the market) is a key enabler of the climate transition. It can take the form of debt, equity, grants, guarantees, technical assistance, or a combination thereof, and can be deployed to de-risk projects through financial innovation, build institutional capacity, and overcome initial cost barriers. Global flows of concessional finance have increased substantially, reaching nearly US$81 billion in 2022 compared to US$54 billion in 2019 (Naran et al. 2024). However, the region remains underserved, having received on average only about 3–4 percent of available concessional climate finance.

The challenge is not solely one of availability. It also reflects limited capacity within public authorities, public banks, and relevant ministries to design and structure transactions that can maximize the mobilization of private capital (including domestic) while minimizing the use of scarce donor or public funds. Strengthening this capacity will be critical to ensure that concessional finance is leveraged effectively—alongside efforts to build a pipeline of bankable projects, improve access to information on climate finance sources, and streamline legal frameworks and application procedures to reduce processing times and administrative hurdles.

Complementary Interventions to Support Green Finance

Several complements are required to scale up financing of green investments throughout MENAAP.

Mobilizing domestic savings. To scale up green investments while shielding them from external shocks associated with foreign currency debt, it is important to mobilize domestic savings. Existing domestic savings in MENAAP are usually invested in foreign assets or abroad rather than within the local economies. In the region, total net foreign assets acquired are estimated at US$2.8 trillion over 2004–23 (Volz et al. 2024).[7] These are investments in foreign economies that could have been invested domestically to support adaptation and mitigation projects.

Mobilizing domestic savings can be encouraged by enhancing financial inclusion; promoting competition (reducing state-owned banks' dominance and improving market contestability); leveraging technology (such as mobile banking and digital payments); and improving regulations (including tax incentives for increased savings) (Arezki and Senbet 2020; World Bank 2021a).

Deepening financial intermediation. To promote financial innovation as well as allocative efficiency, a deepening of financial intermediation is also needed. MENAAP is lagging in these dimensions. Overcoming these issues will require strengthening regulatory frameworks related to credit infrastructure, consumer protection, and transparency in financial transactions while promoting financial literacy, expanding digital payments, and providing targeted support to SMEs by improving both their capabilities and those of banks (IFC 2017).

Developing data infrastructure and disclosure standards. The backbone for these new tools is development of data infrastructure and disclosure standards. Improved data will help structure deals by generating climate risk information and firm-level vulnerability profiles. Sustainability reporting is not yet widespread, but a few countries are moving forward.[8]

Beyond the need to enact appropriate regulations, the lack of internal skills and sustainability expertise remains a major challenge (PwC 2024). Hence, regulators should develop climate-related disclosure standards for firms, have internal units dedicated to climate risks and green finance, and develop practical guidelines so all firms, including SMEs, can implement them.

ENHANCING THE BUSINESS ENVIRONMENT TO SUPPORT ADAPTATION

Although countries in MENAAP all have some form of environmental regulations, their enforcement, scope, and ways of implementation vary greatly.

Climate-Oriented Regulatory Policies

To foster the ability of firms to adapt to the evolution of climate, countries must change a broad range of regulations. Hence, when needed, company laws and firm creation regulations should systematically include harmonized environmental, social, and governance (ESG) standards, and clear sustainability and decarbonization considerations should become part of all existing (or new) investment laws and investment codes. Further, climate regulation should provide clear, simple, and practical standards for energy-efficient, heat-resilient buildings (with stricter provisions for vulnerable locations) as well as decarbonization of production processes.

To curb administrative burdens, licensing and permitting processes should also be streamlined. For example, introducing risk-based approaches to providing construction permits (with a differentiated approach for more or less vulnerable areas) would help firms deal with heat issues. Besides these horizontal interventions, other specific approaches can provide a framework to support adaptation and decarbonization.

Regulatory changes are sometimes bundled within broader green industrial policies. Between 2017 and 2023, industrial policy use increased significantly, with high-income economies driving most of this growth (Ohnsorge et al. 2024). New objectives for industrial policies increasingly include adaptation and the transition to a low-carbon economy.[9] For the region (excluding Afghanistan and Pakistan), the greatest number of industrial policy measures announced over 2018–23 (17 percent of them) concern regulations, according to data from the World Bank's Green Transition Policy Database. Regulations typically focus on green standards and conformity assessments, reporting and disclosure requirements, and auditing. These categories are followed by frameworks and strategies (14 percent) and environmental taxes and fees (12.5 percent).[10]

Keeping in mind, as shown in chapter 1, that heat tends to have a greater negative impact on SMEs and enterprises not connected to global value chains (GVCs), regulatory policies should pay special attention to these categories of firms. The focus should thus be on supporting SMEs and helping firms connect with larger enterprises to be part of GVCs.

Location-Specific Policies

Constraints undermining the private sector's capacity and incentives to adapt and decarbonize are difficult to overcome, even more so for SMEs. Adaptation to heat and other climate challenges, as well as the use of low-carbon technologies, often require significant capital investment to benefit from economies of scale, necessitating collective action to reduce fixed costs. Coordination failures can arise from misaligned incentives—particularly, for example, when SMEs do not own their assets (such as buildings) or face information asymmetries (such as monitoring of emissions or resource efficiency).

Eco-industrial parks (EIPs) can help improve resilience to heat and support decarbonization. With the adequate policy and regulatory framework, EIPs can reduce the financial burden on firms, especially SMEs, by investing in common infrastructure and shared facilities. Among the MENAAP countries, Egypt, Jordan, Morocco, Pakistan, and Tunisia have an extensive network of industrial and investment zones that can be leveraged to facilitate greater resource efficiency, reduce emissions, and promote heat adaptation mechanisms (such as turnkey heat-resistant buildings and adapted cooling solutions). The Tangier automotive cluster in Morocco serves as a prime example of how industrial zones can adopt eco-friendly practices (see box 3.2).

BOX 3.2

Morocco: Green Competitiveness in Industrial Acceleration Zones

The government of Morocco aims to reduce by 53 percent its greenhouse gas emissions by 2035. Industrial zones are associated with significant opportunities to reduce the carbon footprint. Former free trade zones are being redeveloped as industrial acceleration zones (IAZs) to strengthen the export-oriented manufacturing ecosystem, which has been successful in expanding the country's automotive and aerospace industries. IAZs offer tax advantages, allow for local sales, and provide relevant support services (turnkey buildings, security, maintenance, and information technology).

Two pilot IAZs—the Tangier Automotive City (TAC) and the Tangier Free Zone (TFZ)—originally exhibited an average carbon intensity above peers (such as similar zones in Türkiye). TAC, which hosts more than 150 automotive firms, is developing several green projects. These initiatives include renewable energy projects, energy efficiency measures, heat management, and waste management. The TFZ is home to firms in automotive, aeronautics, electronics, textiles, agroindustry, and related services. It is pursuing decarbonization through a green transition strategy, focusing on reducing emissions from key industries and implementing renewable energy sources.

Source: Kechichian and Shin 2025.

Implementing EIPs is complex and depends on country specificities. Typical success factors include an enabling legal framework; demand-driven planning relying on public-private dialogue; effective zone promotion and governance with clear objectives; alignment with relevant energy, environmental, and labor policies (at both the national and international levels); and a credible and efficient monitoring and evaluation mechanism (Kechichian and Shin 2025).

Necessary Conditions: Supporting Market Competition and Competitive Neutrality of SOEs and SOBs

In MENAAP, climate action will require a broad public sector approach as SOEs play a large economic role. First, the state is an owner across many high-emitting sectors. In five countries included in the World Bank's Businesses of the State (BOS) database of firms with direct or indirect state ownership,[11] there were 297 enterprises in carbon-intensive sectors,[12] accounting for about US$41.5 billion in revenues. Second, some of these firms are also vulnerable to changes in climate, especially in infrastructure and network industries. Across the five countries (Egypt, Jordan, Lebanon, Morocco, and Tunisia), 4,858 such firms were operating in climate-vulnerable sectors,[13] accounting for US$124.8 billion in revenues and 826,300 employees (refer to annex 3A, figure 3A.4).

The BOS-listed firms and their (implicit and explicit) advantages can slow adaptation and decarbonization by limiting market mechanisms (World Bank 2023a). When they receive preferential treatment or dominate the market, they may deter private sector investors from engaging, hence weakening productivity and competitiveness. These market distortions can lower the private sector's readiness to adapt, innovate, and invest in low-carbon technologies. In addition, the weak financial capacity of many BOS-listed firms impairs their own ability to engage in and mobilize financing for adaptation and decarbonization.

Reforms are thus needed to support competition and market mechanisms. This need implies setting up measures to foster competition but also including climate objectives in the governance framework for BOS firms. They must also be encouraged to adopt carbon accounting and monitoring systems, publish climate objectives, and facilitate monitoring and oversight. They could further be leveraged to spearhead the adoption of sustainability standards and of climate-related disclosure in annual reports. Demand for low-carbon industrial products could also be enhanced through green procurement by public entities, including BOS firms.

In Morocco, the phosphate industry is leading the way with an ambitious climate strategy. The state-owned OCP (Cherifien Office of Phosphates), a world leader in the phosphate industry, has already reduced its carbon footprint considerably and has set the objective of reaching carbon neutrality by 2040. This is timely given the appearance of CBAM requirements on the EU markets.

FOSTERING GREEN COMPETITIVENESS TO SUSTAIN ACCESS TO EUROPEAN MARKETS

Time is of the essence for countries in the region to adapt to the EU and UK CBAMs and the EU CS3D policies. The EU CBAM has been fully effective since January 1, 2026; the UK CBAM will be effective on January 1, 2027; and the EU CS3D entered into force in July 2024. Other low- and middle-income countries have already initiated preparations to adapt to these policies.[14] Several policies can be considered in response to the EU and UK CBAMs as well as the EU CS3D to mitigate their impact and make the most of market opportunities. These policies can be grouped into two categories: policies to boost competitiveness and the introduction of a direct carbon pricing mechanism.

Boosting Competitiveness Through Compliance with Upcoming Climate Regulations

Governments can help firms boost their "green" competitiveness in three ways: helping firms to comply with new requirements, implementing climate mitigation and trade policies, and reforming fuel subsidies.

Helping Firms to Comply with New Regulations

Strengthening national quality infrastructure—metrology, certification, and accreditation—and raising green and social standards will enable firms to prove compliance with European requirements. The strengthening of carbon certification will enable firms to trace and certify their carbon content so that CBAM payments are based on their actual carbon emissions rather than on higher default values. The strengthening of the quality of infrastructure and the upgrading of green and social standards will reduce the cost for firms of proving compliance with the green and social requirements of the EU's CS3D.

The restructuring of affected sectors to comply with the new requirements and regain their competitiveness in the European markets will require government assistance to firms and workers. This includes assisting sectors in aligning with the environmental and social standards of trading partners and helping workers who lose their jobs in the adjustment to transition to new jobs. Such measures will help speed up the adjustment process and reduce negative impacts.

Implementing Climate Mitigation and Trade Policies

The implementation of climate mitigation policies and trade policies that promote green goods and services (including clean energy sources) will help boost the carbon competitiveness of MENAAP firms in the European market. This involves implementing policies that help meet Nationally Determined Contributions commitments under the Paris Agreement and facilitate the transition to lower greenhouse gas (GHG) emissions, promote renewable energy, and minimize social and environmental harm.

Applying energy policies that support clean energy sources through nondiscriminatory, transparent, time-bound, and monitorable incentives for renewable energy projects would reduce firms' reliance on fossil fuels and increase their carbon competitiveness. Reducing tariffs and nontariff measures, value added tax (VAT), and withholding taxes on green imports of capital goods, intermediate inputs, and services would promote their import, which in turn would enhance carbon and environmental competitiveness of exports (Maliszewska et al. 2025).

Reforming Fuel Subsidies

In addition, pursuing fossil fuel subsidy reforms promotes carbon competitiveness because it encourages efficiency and discourages overconsumption. Fossil fuel subsidies artificially lower prices, typically induce overconsumption, lead to higher carbon-intensive output, and represent a significant fiscal cost, draining resources that could have been used otherwise. In MENAAP (excluding Afghanistan and Pakistan), explicit fossil fuel subsidies (that is, when the retail price is below a fuel's supply cost) were the equivalent of 19 percent of GDP in 2022 (IMF 2025).

A phaseout of subsidies would promote energy efficiency, encourage investment in cleaner production methods, and reduce the carbon intensity of exports. This will help lower CBAM-related payments.

Phasing out subsidies would also generate revenues that can be used to cushion vulnerable groups through targeted support. Reforming fuel subsidies is however a politically sensitive topic.

The political economy around the phaseout of fossil fuel subsidies has been prominent in the region. Similar issues are also associated overall with the introduction of a carbon tax. The various attempts to reform subsidies in countries such as Egypt, Jordan, Tunisia, and the Republic of Yemen (AA 2018; Hamaizia and Moerenhout 2022) illustrate the fiscal necessity but also the political sensitivity of these measures. In these countries, reductions in fuel and electricity subsidies sparked protests, often leading governments to delay or partially reverse reforms.[15] These cases show how subsidy reform and carbon pricing policies are deeply intertwined with social contracts in the region.

Experience suggests that reforms are most effective when presented as part of a comprehensive package, with clear communication and compensatory measures that emphasize fairness and social investment. Morocco stands out as an exception in the region, as it paired energy subsidy reform with a broad communication strategy and direct cash transfers or targeted social programs to protect vulnerable groups.

In contexts where political unrest is likely in specific sectors, second-best approaches may be preferable. "Feebate" mechanisms, for instance, tax inputs based on their carbon content and rebate revenues to firms according to their outputs, thereby aligning incentives without creating major losers. Subsidy reforms operate in a similar way: by raising fuel prices, they effectively increase costs in proportion to consumption; the more fuel a firm or household uses, the more it pays. If revenues from these reforms are then redistributed on the basis of outputs (for example, kilowatt-hours produced in the power sector), efficiency is rewarded and renewable energy is not penalized.

International experience points to a number of successful components of subsidy reform (see box 3.3). They typically include:

- A comprehensive sector reform plan with clear long-term objectives;
- Transparent communication and consultation with stakeholders;
- Phased-in price increases;

- Institutional reforms that depoliticize energy pricing (such as the introduction of automatic pricing mechanisms);
- Improvements in SOE efficiency to reduce the need for producer subsidies; and
- Targeted and efficient measures to protect vulnerable populations.

BOX 3.3

Learning from Successful Fossil Fuel Reforms: Shifting to Social Spending for Equity and Mitigation of Risks of Political Unrest

The Arab Republic of Egypt: Proactive Steps for Fiscal Savings and Social Benefit

In 2014, defying political and economic challenges, Egypt's fossil fuel subsidy reform eased fiscal pressures, with the budget deficit falling from 12.9 percent to 8.1 percent of gross domestic product (GDP) between 2013 and 2019. It has also encouraged greater private investment in clean energy, with solar and wind generation growing almost threefold between 2014 and 2019. Strong government ownership and cross-sectoral coordination underpinned the reforms, supported by transparent communication with the public and international partners on the costs of the subsidies.

Fiscal savings were redirected to strengthen social protection, including expanding food subsidies and doubling benefits under the Social Solidarity Pension, which helped protect vulnerable households. Broader energy sector policies reinforced these gains, most notably the 2015 Electricity Law and 2017 Gas Law, switching fuel toward cleaner and more efficient sources, resolving cross-sectoral arrears, improving supply efficiency, rolling out prepaid meters and digital tracking, and curbing unauthorized fuel sales.

Ecuador: After Failure, Remedies to Reinstate Subsidy Reform

After scrapping subsidies too hastily and facing violent unrest in 2019, the government was forced to reinstall fossil fuel subsidies, which accounted for more than 3 percent of GDP in 2022 (Gentilini et al. 2023). In 2024, the government engaged in another attempt and successfully started to phase out gasoline subsidies. This was met with less resistance than in earlier attempts owing to several factors: favorable international oil prices, cash-back payments to compensate vulnerable groups, and improved public communication. Once subsidies were removed, stable and moderate global oil prices in 2024 limited the price increases for customers.

(continued)

BOX 3.3 ***(continued)***

In addition, the government introduced, in advance, monthly cash transfers to registered eligible recipients in the transport sector affected by fuel price reforms (such as registered taxis, minitrucks, vans, and motorcycles or scooters operating as moto-taxis). For the same actors, retail prices were also capped, with a maximum increase of 5 percent and a maximum decrease of 10 percent compared with international prices. Effective communications redirected vulnerable groups' attention from the loss of the subsidy to the gains from the cash-back payment scheme, therefore preventing the escalation of small protests.

Ghana: Success from Public Education and Buy-In

In 2013, the government eliminated subsidies on gasoline, diesel, kerosene, and liquefied petroleum gas (LPG), amounting to approximately US$1.2 billion or 3.2 percent of GDP (Cooke et al. 2015). The removal was met with low resistance because effective, transparent, and targeted education campaigns involving a wide range of stakeholders (institutions, politicians, and mass media) convinced the public that fossil fuel subsidies were inefficient in protecting poor households. Efforts were made to communicate that removing fossil fuel subsidies could free up funds for critical public investments.

Fiscal savings were spent on eliminating fees for attending primary and secondary schools, allocating extra funds to primary health care in the poorest areas, and expanding access to mass transportation and rural electrification. The reform went further in 2023 by eliminating price caps and allowing for market-based determination of fossil fuel subsidies.

Zambia: Directly Linking Subsidy Reform and Increased Social Protection

In December 2021, Zambia's government removed subsidies on fossil fuels, costing approximately 2.4 percent of GDP (World Bank 2023b). The initial public dissatisfaction due to rapid fossil fuel price hikes was eased thanks to the government's commitment to replacing inefficient fossil fuel subsidies with an increase in social protection spending, raising the latter's (target) share in GDP from 0.7 percent in 2020 to 1.6 percent by 2025 (World Bank 2023c). The expansion of social programs increased the number of recipients as well as the amount of cash transfers, supported free schooling, and enabled hiring of more teachers and health workers (mainly in rural areas). Lower fiscal outlays from phasing out fossil fuel subsidies also allowed the Zambian government to reverse years of lapses in budget sustainability.

Sources: Hallegatte et al. 2023; World Bank 2023b.

Direct Carbon Pricing to Retain Carbon Revenue and Reduce Emissions Intensity

Countries in the region can retain CBAM-related revenues domestically by establishing a *direct* carbon pricing mechanism, ensuring that payments remain within the national economy rather than being transferred overseas. The CBAM payments could be reduced by deducting domestic direct carbon price payments to MENAAP governments should such a carbon price be established.

Direct carbon pricing refers to carbon pricing instruments that apply a price incentive directly in proportion to the GHG emissions generated by a given product or activity. The primary forms of direct carbon pricing are a carbon tax and the EU's Emissions Trading System (ETS).

Advantages to Offsetting CBAM Costs

Governments can leverage existing taxes to introduce a direct carbon price instrument. Because carbon emissions are proportional to fuels used for combustion, a tax on fuel consumption can effectively serve as a carbon price. Both carbon taxes and the ETS qualify for credit under CBAM, but an ETS requires years to implement fully, whereas carbon taxes can be more easily administered through existing tax systems, allowing for an immediate response.

Leveraging existing taxes for a swift response. Because most MENAAP countries already tax carbon indirectly through fuel excise taxes, stamp duties, and development taxes, reclassifying these charges into a direct carbon tax could present a first step to respond to CBAM. Strengthening the existing excise tax system to include an explicit carbon tax on all fuels, including natural gas and coal, would cover fuel-related emissions and a large share of CBAM financial obligations while keeping administrative and compliance costs low. An upstream carbon tax administration implies few points of regulation, limiting the number of entities responsible for collecting and delivering the tax due to the government.

Given CBAM's phased implementation, a gradual transition to a domestic carbon fuel tax would allow a swift response to CBAM by covering emissions from fuel combustion and provide time for adaptation. For instance, Argentina and Uruguay introduced a carbon tax in March 2018 and January 2022, respectively, by converting their indirect carbon pricing to direct carbon pricing for specific fuels (Rabbia 2023).

Public revenue generation. Importantly, carbon emissions represent a significantly underutilized tax base that could boost public revenues. Unlike traditional tax bases, such as labor or corporate income, taxing pollution can enhance economic efficiency by correcting market failures—internalizing negative externalities associated with air pollution rather than discouraging productive activities.

Detailed computations performed for Egypt show the promises of this approach. Figure 3.1 indicates that in the case of Egypt, gradually implementing a broad-based carbon tax, when aligned with the carbon price needed to retain revenues from the EU CBAM (that is, to offset the CBAM certificate costs for energy-related emissions),[16] could generate revenues equivalent to approximately 3 percent of GDP by 2034, when the EU CBAM is fully phased in. If Egypt also decides to further phase out fossil fuel subsidies, as they dilute the carbon price signal, revenues could increase by an additional percentage point of GDP in 2034, representing more than 4 percent of GDP.

FIGURE 3.1

Potential Revenue and GDP Increases from Carbon Pricing and Taxation in Egypt, Arab Rep., 2026–36

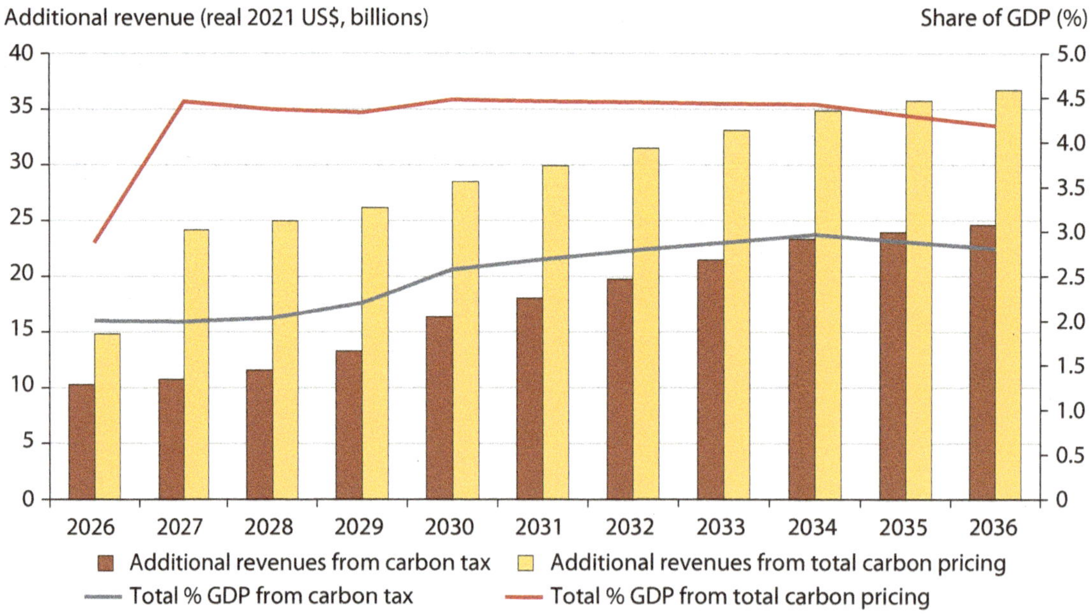

Sources: World Bank calculations based on the Climate Policy Assessment Tool (CPAT) of the International Monetary Fund and World Bank Group (Black et al. 2023).

Note: The figure illustrates projected additional revenues from carbon pricing policies in Egypt from 2026 to 2036. The "carbon tax" scenario sets carbon prices at levels required to offset Carbon Border Adjustment Mechanism (CBAM) obligations without accounting for the subsidies. The "total carbon pricing" scenario builds on the carbon tax approach by also eliminating fossil fuel subsidies, price controls, and value added tax (VAT) differentials. Both scenarios factor in CBAM's gradual implementation. GDP = gross domestic product.

Productivity gains. Evidence from emerging economies shows that energy subsidy reforms can generate productivity gains. In Indonesia and Mexico, a 10 percent increase in fuel prices was associated with a rise in total factor productivity of 1.4 percent and 1.6 percent, respectively, as firms shifted from inefficient fuel-powered equipment toward more productive, electricity-powered technologies (Calì et al. 2022). Similarly, in Oman, a 10 percent increase in fossil fuel prices raised labor productivity by 4.1–4.6 percent and boosted total factor productivity even more strongly, reflecting firm-level efficiency gains and investment in information and communication technology (ICT) and production upgrading (Amann et al. 2021).

Jobs and equity benefits. The benefits of carbon taxation extend beyond environmental goals. An economywide carbon tax on fuels can yield significant benefits, including job creation and greater equity, while fostering a simpler, more growth-friendly system that extends to the informal sector. Upstream carbon taxation can have lower costs for economic growth than conventional taxes. Covering large informal sectors and using the revenues to substitute formal-sector taxes creates opportunities to boost output. A broad tax base allows carbon taxation to raise revenues more efficiently, with less negative impact on output than conventional taxes.[17]

Employment outcomes are also likely better with upstream carbon taxes than with some conventional taxes, especially those on labor (Elmaleh et al. 2021; Metcalf and Stock 2023; Schoder 2023; World Bank 2022d). At the macro level, the literature shows, carbon pricing can shift resources from capital-intensive sectors like industry to more labor-intensive ones such as services, while also promoting low-carbon, labor-intensive production methods. This increases demand for labor, raising wages relative to capital returns and enhancing equity, as capital is mostly owned by wealthier individuals (Goulder et al. 2019; Hille and Möbius 2019; Malerba and Wiebe 2021; Moszoro 2024; Saget et al. 2020). However, at the micro level, workers in carbon-intensive sectors will need targeted support and transition measures.

Other development effects. For example, development co-benefits are particularly pronounced in Egypt. By discouraging pollution-intensive activities, carbon taxation can improve air quality, enhance public health, and reduce social costs that come with road use, including congestion, accidents, and road damage. These effects could be so large that many researchers have found they may exceed the economic cost of mitigation

(Markandya et al. 2013; Rauner et al. 2020; Scovronick et al. 2019; Vandyck et al. 2018; Weitzel et al. 2019).

All this means that there are national economic interests for taxing carbon, independent from its contribution to climate objectives and reduced CBAM liabilities (Heine and Black 2019).

Upstream and Downstream Carbon Taxes

Aligning the domestic carbon price with the full coverage of the CBAM levy can be done in two steps: introducing a carbon tax within an existing excise tax system and introducing a complementary downstream carbon tax.

Using an existing excise tax system to introduce an explicit carbon tax on all fuels, including natural gas and coal, would cover fuel-related emissions. Standard values of emissions from combustion can be used to calculate the tax rates. Such a revised tax legislation should thus ensure the carbon fuel excise tax applies only to fuels used for combustion. Fuels like gasoline and diesel are primarily motor fuels, thus noncombustion uses are negligible. However, energy products such as coal, natural gas, and heavy fuel oil can be used in industrial processes by means other than combustion. Small quantities can serve as feedstock into industrial CBAM processes, which do not emit GHGs.[18] Coal, natural gas, and heavy fuel oil can also give rise to emissions from certain industrial processes other than combustion.

A second step of a reform could consist of introducing a separate downstream carbon tax to cover emissions from noncombustion industrial processes and product use (IPPU),[19] which would complement the upstream carbon-based tax. This would ensure that the carbon price instrument covers all CBAM emissions. The downstream tax would be paid by operators of the installations engaged in industrial production of goods, while the primary carbon tax on combustion of fuels would be collected by fuel distributors. To avoid complexity, administrative thresholds could be used.

ECONOMIC OPPORTUNITIES ARISING FROM MORE SUSTAINABLE ECONOMIES

The necessary move to more sustainable economies will bring new economic opportunities. Firms in the region should be able to seize them if the previous key policy changes have been implemented.

Current research, from this report and also from the World Bank's CCDRs, suggests the existence of opportunities linked to heat adaptation as well as to the new trade measures in Europe that will offer prospects for some countries in the region. Beyond these possibilities, the development of the renewable energy market as well as the important role of resource efficiency for the region will offer further business opportunities.

Opportunities Linked to Heat Adaptation

Adapting to climate change will require the development of products and services tailored to the specifics of the region. In this respect, sustainable cooling solutions are a new critical market as temperatures increase and extreme heat events become more frequent. Cooling is vital for maintaining or enhancing productivity, preserving perishable goods, and supporting health care services, thereby also benefiting vulnerable populations.

Cooling solutions must be sustainable,[20] which implies a need to maximize reliance on so-called *passive* or *nature-based* strategies such as:

- Using materials such as cool roofs, high-albedo walls, shading, and insulation to keep heat out (World Bank 2022e, 2022f)
- Fostering urban infrastructure like parks for shade and water features to reduce ambient temperatures (Lee et al. 2024; World Bank 2022f)
- Promoting climate- and tradition-informed design involving crossventilation or courtyards (World Bank 2025e).

A second option relies on *active technological solutions*—prioritizing energy-efficient air conditioning, implementing energy performance standards, and fostering a refrigerant transition to options with low or medium global warming potential (Ayres et al. 2020; ESMAP 2023).[21]

This will create significant business opportunities for investors, involving horticulture, construction, and manufacturing as well as consulting and engineering. The global annual demand for cooling solutions is projected to reach at least US$600 billion by 2050 (UNEP and IFC 2024). This growth is largely driven by active cooling solutions, while passive strategies currently account for about 10 percent of the market. The estimated market opportunity (figure 3.2) for the Middle East is US$46 billion by 2050 against US$20 billion in 2023 (UNEP and IFC 2024).

FIGURE 3.2

Estimated Evolution of the Sustainable Cooling Market, by Region, 2023–50

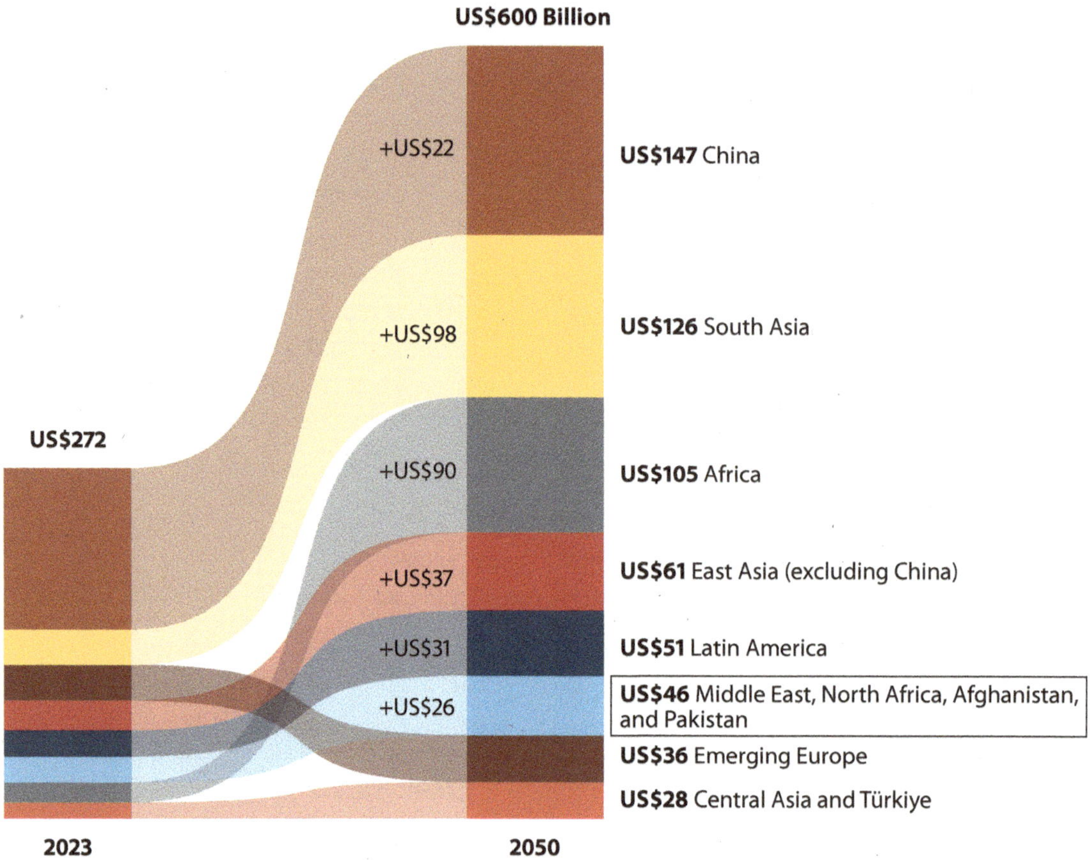

Source: UNEP and IFC 2024.

Opportunities Generated by European Trade Policies

European climate mitigation policies on trade and foreign direct investment (FDI) inflows will also offer opportunities for firms in the region. Countries whose exports have a lower carbon intensity than the EU average or the UK average will be in a good position to benefit from the introduction of the EU CBAM or UK CBAM, respectively. Countries such as Jordan and Morocco have lower carbon intensity in aluminum production than the EU average (chapter 2), and countries such as Oman and Qatar have lower carbon intensity in aluminum production than the UK average (refer to chapter 2, annex 2A). Thus, they may relatively quickly gain competitiveness in those markets, maintain market access, and potentially increase their market share if they can upgrade their quality infrastructure to be able to reliably certify the carbon intensity of their exports.

For countries with a carbon emission intensity higher than the EU average (for instance, Egypt in aluminum or Tunisia in cement) or higher than the UK average (for instance, Egypt in aluminum and fertilizer and Tunisia in iron and steel), eventually regaining access to the European markets will be a lengthier process. It will require not only upgrades to quality infrastructure but also improvement of production processes, a long and potentially costly endeavor.

The EU CS3D also opens opportunities for compliant MENAAP firms to remain in EU GVCs and possibly expand their market share, to the detriment of firms from other countries that do not. Firms that comply with the rigorous environmental and human rights due diligence requirements could see market opportunities to expand in EU GVCs like agriculture, chemical and nonmetallic minerals, metal products, mining and quarrying, petroleum, and textile and apparel, among others.

Opportunities Linked to the Energy Sector

The development of clean energy sources will support green competitiveness in EU markets and will also have an impact on jobs in the region. MENAAP has a key strength for renewable energy development: its abundant natural resources. The region benefits from some of the world's highest solar irradiance. Several countries—Jordan, Oman, Saudi Arabia, and the Republic of Yemen—indeed have solar photovoltaic (PV) capacity factors above 25 percent.[22] Further, regional wind strengths are high. Countries such as Bahrain, Egypt, Jordan, Oman, and Saudi Arabia have a wind capacity factor above 30 percent (World Bank 2021b).[23] Pakistan also has large offshore wind resources with a potential for 11–21 gigawatts of offshore wind power generation (World Bank 2024b).

Governments in the region are also proactive. For example, in Morocco, government clean energy targets are expected to yield 25,000 new job opportunities per year (World Bank 2022c), while the Egyptian government's targets should lead to 67,000 new job opportunities per year (World Bank 2022a, 2022b). The building, operation, and maintenance of solar power plants (as well as wind farms) will generate direct economic activity in construction, metalworks, mechanical industries, engineering, and consulting.

The required grid modernization, along with energy storage solutions, will enable better integration of intermittent renewable energy and, again, benefit sectors such as construction, metalworks, mechanical industries, engineering, and consulting.

Some countries in the region, like Morocco or Tunisia, may also be able to enter the battery production GVC. Indeed, Tunisia's strength in related industries—a relatively high share of its GDP comes from value added within the electrical fittings and battery sectors—may provide a basis for battery manufacturing in the future (IEA 2024).

Opportunities Linked to Resource Efficiency

Efficient use of resources is a cornerstone of firm-level productivity. Given the MENAAP region's wide variation in firm-level energy efficiency as well as the positive link between energy efficiency and productivity (as discussed in chapter 1), it will be critical to limit energy grid inefficiencies and losses, use energy-efficient production processes, and develop energy-efficient buildings (or retrofit existing ones)—all in compliance with clearly defined energy efficiency standards. This move toward efficiency will open multiple economic opportunities, involving sectors such as agriculture, transport, manufacturing, consulting, and engineering. A good example of the current movement toward green and efficient production processes is the 2024 agreement between Sonatrach and Tosyali Algérie (a subsidiary of Tosyali Holding, a Turkish company) to undertake a feasibility study to define how to produce hydrogen from renewable energies (Veysel 2024).

Accompanying technological changes in adjacent sectors will also open opportunities. For example, the shift toward electric vehicles (EVs) implies that battery production for cars is now a promising business that involves chemical and electrical industries, component part manufacturing, construction, transportation, and energy. Although countries in the region are mostly absent from the upstream (raw materials extraction) part of the value chain, there may be opportunities in the midstream segment (with the processing of the raw materials and the creation of anodes and cathodes) as well as the downstream segment, which involves the assembly of the battery cells into modules, then packing and selling them to automakers (Carreon 2023).

For countries with limited experience in the field, FDI and technology transfers are good ways to enter the market. Morocco is a case in point. "Gotion Power Morocco," a subsidiary of the Chinese-European EV battery producer "Gotion High Tech," is about to build a gigafactory in northwest Morocco for a total expected investment of US$6.5 billion (Faouzi 2025). As currently planned, the first phase of the investment will cost US$1.3 billion. It will focus on the production of batteries and anodes and cathodes to be exported to Europe.

NOTES

1. The Falkenmark "absolute water scarcity" threshold applies to a country with less than 500 cubic meters of renewable freshwater resources per person per year.
2. For individual country reports, see https://www.worldbank.org/en/publication/country-climate-development-reports.
3. SMEs in MENAAP receive a limited amount of banks' credit. SME financing is mostly secured by mortgages on real estate, with high collateral-to-loan value ratios (refer to annex 3A, figure 3A.2)—from 127 percent in Jordan up to 281 percent in the Republic of Yemen. Women-led SMEs also face challenges. Among the 23 percent of micro, small, and medium enterprises (MSMEs) that are led by women in the region, 55 percent lack access to credit financing, and only 4 percent declare being well served by financial institutions (Dornel et al. 2020). Further, SMEs have limited access to specialized trade finance instruments and face constraints because of Anti-Money Laundering (AML) and Know-Your-Customer (KYC) requirements. All of this prevents SMEs from investing significantly in adaptation and mitigation.
4. For example, an analysis of commercial loans provided to private firms in Mexico shows that, following a temperature shock, affected firms faced increased challenges in accessing credit, paid higher interest rates, and provided more collateral (Aguilar-Gomez et al. 2024). Similarly, negative precipitation shocks in the region are negatively correlated with having a loan or line of credit (Gatti et al. 2024; Zaveri et al. 2024).
5. Green finance is any structured financial activity created to ensure a better environmental outcome. It includes an array of loans, debt mechanisms, and investments that are used to encourage the development of green projects, minimize the impact on the climate of more regular projects, or both (Fleming 2020).
6. Green bond league tables (updated quarterly or annually by financial institutions) rank issuers based on the volume of green bonds they issue, helping investors to identify leading entities in sustainable finance and to promote transparency and accountability in green financing. Key metrics include total issuance, the number of deals, and market share.
7. The MENAAP estimate excludes Afghanistan and Pakistan.

8. In the Gulf Cooperation Council (GCC) countries, although sustainability reporting by listed companies remains low overall, reporting rates have improved in recent years (Farooq et al. 2021). More recently, the Central Bank of Jordan issued its first Climate Risk Management Regulations (CBJ 2025) in 2025, which will require financial institutions to improve climate data and disclosure standards. In Morocco, the Central Bank also issued in early 2025 a directive for financial institutions on the collection and reporting of climate exposure data (*Finances News Hebdo* 2025).
9. Such policies make up 28 percent of the entries in the New Industrial Policy Observatory (NIPO) database worldwide—each NIPO entry refers to a distinct state intervention. The NIPO dataset used for this report records policies that were implemented or announced on or after January 1, 2023. The monitoring covers 75 jurisdictions making up 94 percent of global GDP. Of those, 10 jurisdictions are in MENAAP (including Pakistan but excluding Afghanistan), representing 13 percent of the database (Evenett et al. 2024).
10. The Green Transition Policy Database harmonizes secondary data using a common taxonomy to provide insights into the global use of climate policies, both mitigation and adaptation, from 2018 to 2023.
11. The World Bank's global "Businesses of the State" (BOS) database of firms with state participation was developed for the Bank's 2023 report, *The Business of the State* (World Bank 2023a). It includes 76,400 firms with direct or indirect state ownership of at least 10 percent across 91 countries, and 18 percent are in high-emitting sectors. The database includes five countries of relevance for this report: Egypt, Jordan, Lebanon, Morocco, and Tunisia. Carbon-intensive BOS-listed firms in the region operate mainly in the transportation, energy, extractive, and manufacturing sectors.
12. Carbon-intensive sectors include energy, transportation, financial services, agriculture, and raw materials. High-emitting sectors were identified using the World Resource Institute's "World Greenhouse Gas Emissions: 2019" (https://www.wri.org/data/world-greenhouse-gas-emissions-2019) and converted into Statistical Classification of Economic Activities in the European Community (NACE) code.
13. Sectors prone to the negative effects of climate change were identified using a risk classification by sectors and converted into NACE code. The risk classification was corroborated with results from the literature.
14. Many low- and middle-income countries are already preparing for the EU CBAM and EU CS3D (Krajcovicova 2025). They include India, Türkiye, and Viet Nam. For further discussion of the steps these countries have taken, see annex 3B.
15. Fuel subsidy reforms in several MENA countries have faced disruption from public protests. In Tunisia (2013–14), demonstrations constrained implementation and slowed reform momentum. In the Republic of Yemen (2014), fuel price increases contributed to widespread unrest that preceded the Houthi takeover of Sana'a, effectively derailing reform. And in Jordan (2018), anti-austerity protests linked to energy price measures temporarily disrupted reform efforts (Hamaizia and Moerenhout 2022).

16. In Egypt, the carbon price needed to offset CBAM obligations for energy-related emissions depends on the highest estimated carbon price calculated across the CBAM sectors according to the export volume, Egyptian emission intensities, EU benchmark emission intensities, the EU ETS price, and the CBAM phase-in factor. In Egypt, the highest carbon price is estimated to be for fertilizer products.
17. See, for example, evidence for China and India (Carson et al. 2021); Côte d'Ivoire (Timilsina et al. 2024); Eastern Europe (World Bank 2022d, 48f); the EU (Barrios et al. 2013); Ghana (Elmaleh et al. 2021, 154f); the Islamic Republic of Iran (Mirhosseini et al. 2017); Pakistan (Burns et al. 2021); Spain (Markandya et al. 2013); and globally (Bento et al. 2018; Schoder 2023).
18. For example, natural gas is used as a feedstock in the production of chemicals and plastics. Crude oil can be used to produce naphtha, which is then used to produce plastic, but the oil is not burned and thus has not emitted.
19. IPPU emissions can be large in certain sectors, such as production of cement and fertilizers.
20. Conventional cooling solutions such as more air conditioners and refrigeration systems imply a higher demand for power, which in itself triggers an increase in fossil fuel emissions under the current generation system, thus warming the climate further.
21. Global warming potential (GWP) allows comparisons of the global warming impacts of different gases. It is a measure of how much energy the emission of 1 ton of a gas will absorb over a given period of time relative to the emission of 1 ton of CO_2. The larger the GWP, the more that a given gas warms the Earth compared with CO_2 over that time period (EPA 2025).
22. The "solar PV capacity factor" is the ratio of a solar plant's actual energy output over a specific period of time to its maximum possible output if it were running at full power permanently over the same period of time.
23. The "wind capacity factor" is the ratio of a wind turbine's actual energy produced over a specific period of time to its maximum possible output if it ran at full power permanently over the same period of time.

REFERENCES

AA (Andalou Agency). 2018. "Reforming Energy Subsidies in the MENA." Analysis, April 16. https://www.aa.com.tr/en/energy/analysis/reforming-energy-subsidies-in-the-mena-/19673.

Acemoglu, D., P. Aghion, L. Bursztyn, and D. Hémous. 2012. "The Environment and Directed Technical Change." *American Economic Review* 102 (1): 131–66.

Adam, B., S. Hallegatte, and O. Mahul. 2025. "Mobilizing Public-Private Solutions to Manage the Financial Impacts of Natural Hazards in Emerging Markets and Developing Economies: Challenges and Opportunities of Operationalizing Public-Private Insurance Programs." Prosperity Note 200025, World Bank.

Aghion, P., A. Dechezleprêtre, D. Hémous, R. Martin, and J. Van Reenen. 2016. "Carbon Taxes, Path Dependency, and Directed Technical Change: Evidence from the Auto Industry." *Journal of Political Economy* 124 (1): 1–50.

Aguilar-Gomez, S., E. Gutiérrez, D. Heres, D. Jaume, and M. Tobal. 2024. "Thermal Stress and Financial Distress: Extreme Temperatures and Firms' Loan Defaults in Mexico." *Journal of Development Economics* 168 (3): 103246. doi:10.1016/j.deveco.2023.103246.

Amann, J., N. Cantore, M. Calì, V. Todorov, and C. F. C. Cheng. 2021. "Switching It Up: The Effect of Energy Price Reforms in Oman." *World Development* 142: 105252.

Arezki, R., and L. W. Senbet. 2020. "Transforming Finance in the Middle East and North Africa." Policy Research Working Paper 9301, World Bank.

Ayres, M., N. Stankevich, and A. Diehl. 2020. "Mobile Cooling: Assessment of Challenges and Options." Background paper, World Bank.

Barrios, S., J. Pycroft, and B. Saveyn. 2013. "The Marginal Cost of Public Funds in the EU: The Case of Labour Versus Green Taxes." In *Fiscal Policy and Growth*, 403–31. Paper presented at the Banca d'Italia workshop, "Fiscal Policy and Growth," Perugia, Italy, March 29–31, 2012.

Bento, A. M., M. R. Jacobsen, and A. A. Liu. 2018. "Environmental Policy in the Presence of an Informal Sector." *Journal of Environmental Economics and Management* 90: 61–77.

Black, S., I. Parry, V. Mylonas, N. Vernon, and K. Zhunussova. 2023. "The IMF–World Bank Climate Policy Assessment Tool (CPAT): A Model to Help Countries Mitigate Climate Change." Working Paper 2023/128, International Monetary Fund.

Burns, A., C. Jooste, and G. Schwerhoff. 2021. "Climate Modeling for Macroeconomic Policy: A Case Study for Pakistan." Policy Research Working Paper 9780, World Bank.

Calì, M., N. Cantore, L. Iacovone, M. Pereira-López, and G. Presidente. 2022. "Too Much Energy: The Perverse Effect of Low Fuel Prices on Firms." *Journal of Environmental Economics and Management* 111: 102587.

Carreon, A. R. 2023. "The EV Battery Supply Chain Explained." Rocky Mountain Institute, May 5. https://rmi.org/the-ev-battery-supply-chain-explained/.

Carson, R. T., M. R. Jacobsen, and A. A. Liu. 2021. "Are Carbon Taxes More Efficient in Industrializing Countries? Comparing China and India to the United States." Working paper, Center for Environmental Economics, University of California, San Diego. https://econweb.ucsd.edu/cee/workingpapers.html.

CBJ (Central Bank of Jordan). 2025. "Climate Risk Management Regulations, No. (2) of 2025." Ref. no. 23/1/3551, CBJ.

Cirera, X., D. Comin, and M. Cruz. 2022. *Bridging the Technological Divide: Technology Adoption by Firms in Developing Countries*. World Bank.

Cooke, E. F. A., S. Hague, L. Tiberti, J. Cockburn, and A.-R. El Lahga. 2015. "Estimating the Impact on Poverty of Ghana's Fuel Subsidy Reform and a Mitigating Response." *Journal of Development Effectiveness* 8 (1) 105–28. doi:10.1080/19439342.2015.1064148.

Dornel, A. D., M. Ait Ali Slimane, and K. Mohindra. 2020. "Improving SME Access to Trade Credit and Financing in MENA." MENA Knowledge and Learning Quick Note 180, World Bank.

Elmaleh, D. J., K. A. Stapleton, M. A. Dutz, E. G. Graham, A. Kruse, and K. G. Kwakye. 2021. "Ghana Rising: Accelerating Economic Transformation and Creating Jobs." Ghana Country Economic Memorandum, Report AUS00002590, World Bank.

Emirates News Agency. 2024. "UAE Tops Regional Green Bond League Tables with $10.7 Billion." Emirates News Agency, January 25. https://www.wam.ae/en/article/b1bvln9-uae-tops-regional-green-bond-league-tables-with.

EPA (US Environmental Protection Agency). 2025. "Understanding Global Warming Potentials." Topic page, EPA. https://www.epa.gov/ghgemissions/understanding-global-warming-potentials.

ESMAP (Energy Sector Management Assistance Program). 2023. "Energy Sector Management Assistance Program Annual Report 2022." World Bank.

Evenett, S., A. Jakubik, F. Martín, and M. Ruta. 2024. "The Return of Industrial Policy in Data." *The World Economy* 47 (7): 2762–88.

Faouzi, A. 2025. "Gotion Gigafactory Construction Launches Within Days in Kenitra." Morocco World News, May 22. https://www.moroccoworldnews.com/2025/05/202536/gotion-gigafactory-construction-launches-within-days-in-kenitra.

Farooq, M. B., R. Zaman, D. Sarraj, and F. Khalid. 2021. "Examining the Extent of and Drivers for Materiality Assessment Disclosures in Sustainability Reports." *Sustainability Accounting, Management and Policy Journal* 12 (5): 965–1002. doi:10.1108/SAMPJ-04-2020-0113.

Finances News Hebdo. 2025. "Banking Sector: Bank Al-Maghrib Publishes a New Directive on Climate Risks." *Finances News Hebdo*, January 24. https://fnh.ma/article/actualite-economique/secteur-bancaire-bam-maroc-climat-risque.

Flammer, C. 2021. "Corporate Green Bonds." *Journal of Financial Economics* 142 (2): 499–516.

Fleming, S. 2020. "What Is Green Finance and Why Is It Important?" Article for the World Economic Forum "Green Horizon Summit: The Pivotal Role of Finance," Cologny, Switzerland, November 9–11. https://www.weforum.org/stories/2020/11/what-is-green-finance/.

Gatti, R., A. Islam, C. Maue, and E. Zaveri. 2024. "Thirsty Business: A Global Analysis of Extreme Weather Shocks on Firms." Policy Research Working Paper 10923, World Bank.

Gatti, R., H. Onder, A. Islam, et al. 2025. *Shifting Gears: The Private Sector as an Engine for Economic Growth in the Middle East and North Africa*. MENA Economic Update, April, World Bank.

Gentilini, U., M. B. Alsafi Almenfi, H. TMM Iyengar, et al. 2023. "Tracking Global Social Protection Responses to Inflation." Living Paper v. 5, World Bank.

Goulder, L. H., M. A. Hafstead, G. Kim, and X. Long. 2019. "Impacts of a Carbon Tax Across US Household Income Groups: What Are the

Equity-Efficiency Trade-Offs?" *Journal of Public Economics* 175 (1): 44–64.

Hallegatte, S., C. Godinho, J. Rentschler, et al. 2023. *Within Reach: Navigating the Political Economy of Decarbonization*. Climate Change and Development Series. World Bank.

Hamaizia, A., and T. Moerenhout. 2022. "Five Takeaways from a Decade of Energy Subsidy Reforms in MENA." Chatham House, February 16. https://www.chathamhouse.org/2022/02/five-takeaways-decade-energy-subsidy-reforms-mena.

Heine, D., and S. Black. 2019. "Benefits Beyond Climate: Environmental Tax Reform." In *Fiscal Policies for Development and Climate Action*, edited by M. A. Pigato. World Bank.

Hille, E., and P. Möbius. 2019. "Environmental Policy, Innovation, and Productivity Growth: Controlling the Effects of Regulation and Endogeneity." *Environmental and Resource Economics* 73 (4): 1315–55.

IEA (International Energy Agency). 2024. "Energy Technology Perspectives 2024." Report, IEA.

IFC (International Finance Corporation). 2017. "MSME Finance Gap: Assessment of the Shortfalls and Opportunities in Financing Micro, Small and Medium Enterprises in Emerging Markets." Report 121264, IFC, World Bank Group.

IMF (International Monetary Fund). 2025. "Climate Change: Fossil Fuel Subsidies." Topical overview, web page, IMF. https://www.imf.org/en/Topics/climate-change/energy-subsidies#:~:text=Explicit%20subsidies%20are%20broadly%20found,all%20regions%20except%20North%20America.

IRENA (International Renewable Energy Agency). 2016. *Unlocking Renewable Energy Investment: The Role of Risk Mitigation and Structured Finance*. IRENA.

Jaffe, A. B., R. G. Newell, and R. Stavins. 2004. "A Tale of Two Market Failures: Technology and Environmental Policy." *Ecological Economics* 54 (2–3): 164–74.

Kechichian, E. R., and N.-Y. Shin. 2025. "Leveraging Eco-Industrial Parks for a Sustainable Transition of Industries." Report 201201, World Bank.

Kling, G., U. Volz, V. Murinde, and S. Ayas. 2021. "The Impact of Climate Vulnerability on Firms' Cost of Capital and Access to Finance." *World Development* 137: 105131. doi:10.1016/j.worlddev.2020.105131.

Krajcovicova, J. 2025. "Implications of New Environmental Regulations." Business Environment Working Group, Donor Committee for Enterprise Development.

Lee, H., J. Hasoloan, H. Park, T. B. Chapman, and J. Siri. 2024. "Combating Heat in Cities: Operationalizing the Urban Heat Agenda at the World Bank." Report 194295, World Bank.

Malerba, D., and K. S. Wiebe. 2021. "Analysing the Effect of Climate Policies on Poverty Through Employment Channels." *Environmental Research Letters* 16 (3): 035013.

Maliszewska, M., C. Fischer, E. Jung, and M. Chepeliev. 2025. "Carbon Border Adjustment Mechanism (CBAM) Exposure Indices Methodological Note." Prosperity Note 202614, World Bank.

Markandya, A., M. González-Eguino, and M. Escapa. 2013. "From Shadow to Green: Linking Environmental Fiscal Reforms and the Informal Economy." *Energy Economics* 40 (Suppl 1): S108–S118.

Metcalf, G. E., and J. H. Stock. 2023. "The Macroeconomic Impact of Europe's Carbon Taxes." *American Economic Journal: Macroeconomics* 15 (3): 265–86.

Mirhosseini, S. S., N. Mahmoudi, and S. N. Pourali Valokolaie. 2017. "Investigating the Relationship Between Green Tax Reforms and Shadow Economy Using a CGE Model—A Case Study in Iran." *Iranian Economic Review* 21 (1): 153–67.

Moszoro, M. W. 2024. "The Direct Employment Impact of Public Investment." *International Journal of Management and Economics* 60 (1): 59–74.

Naran, B., T. Zhang, and I. Gupta. 2024. *Understanding Global Concessional Climate Finance 2024.* Climate Policy Initiative. https://www.climatepolicyinitiative.org/publication/understanding-global-concessional-climate-finance-2024/.

OECD (Organisation for Economic Co-operation and Development). 2022. "OECD Blended Finance Guidance for Clean Energy." OECD Environment Policy Paper 31.

Ohnsorge, F., M. Raiser, and Z. Leiyu Xie. 2024. "The Renaissance of Industrial Policy: Known Knowns, Known Unknowns, and Unknown Unknowns." *Let's Talk Development* (blog), October 29, World Bank. https://blogs.worldbank.org/en/developmenttalk/the-renaissance-of-industrial-policy--known-knowns--known-unknow.

PwC (PricewaterhouseCoopers). 2024. "Sustainability in the Middle East 2024: Breaking Through Barriers." Third PwC Sustainability Middle East Survey Report, June 5. https://www.pwc.com/m1/en/sustainability/insights/sustainability-in-the-middle-east-2024.html.

Rabbia, M. 2023. "Why Did Argentina and Uruguay Decide to Pursue a Carbon Tax? Fiscal Reforms and Explicit Carbon Prices." *Review of Policy Research* 40 (2): 230–59. doi:10.1111/ropr.12517.

Rauner, S., J. Hilaire, D. Klein, J. Strefler, and G. Luderer. 2020. "Air Quality Co-benefits of Ratcheting Up the NDCs." *Climatic Change* 163 (3): 1481–1500.

Saget, C., A. Vogt-Schilb, and T. Luu 2020. *Jobs in a Net-Zero Emissions Future in Latin America and the Caribbean.* Inter-American Development Bank and International Labour Organization.

Schoder, C. 2023. "Regime-Dependent Environmental Tax Multipliers: Evidence from 75 Countries." *Journal of Environmental Economics and Policy* 12 (2): 124–67.

Scovronick, N., M. Budolfson, F. Dennig, et al. 2019. "The Impact of Human Health Co-benefits on Evaluations of Global Climate Policy." *Nature Communications* 10 (1): 2095.

Tang, D. Y., and Y. Zhang. 2020. "Do Shareholders Benefit from Green Bonds?" *Journal of Corporate Finance* 61: 101427.

Timilsina, G. R., Y. Dissou, M. Toman, and D. Heine. 2024. "How Can a Carbon Tax Benefit Developing Economies with Informality? A CGE Analysis for Côte d'Ivoire." *Climate Policy* 24 (1): 71–86.

UNEP and IFC (United Nations Environment Programme and International Finance Corporation). 2024. *Cooler Finance: Mobilizing Investment for the Developing World's Sustainable Cooling Needs*. UNEP and IFC.

Vandyck, T., K. Keramidas, A. Kitous, et al. 2018. "Air Quality Co-benefits for Human Health and Agriculture Counterbalance Costs to Meet Paris Agreement Pledges." *Nature Communications* 9: 4939. doi:10.1038/s41467-018-06885-9.

Veysel, K. 2024. "Sonatrach and Tosyali Sign MoU to Explore Green Hydrogen Production in Algeria." The Net-ZeroCircle, August 1.

Volz, U., Y. Lo, and V. Mishra. 2024. "Scaling Up Green Investment in the Global South: Strengthening Domestic Financial Resource Mobilisation and Attracting Patient International Capital." Technical report, SOAS Centre for Sustainable Finance, University of London. doi:10.25501/SOAS.00041078.

Wang, J., X. Chen, X. Li, J. Yu, and R. Zhong. 2020. "The Market Reaction to Green Bond Issuance: Evidence from China." *Pacific-Basin Finance Journal* 60: 101294.

Weitzel, M., T. Vandyck, K. Keramidas, et al. 2019. "Model-Based Assessments for Long-Term Climate Strategies." *Nature Climate Change* 9 (5): 345–47.

World Bank. 2021a. "Middle East and North Africa (MENA) Regional Update 2021: Preventing a Lost Decade, Unlocking the Region's Potential." PowerPoint presentation, World Bank.

World Bank. 2021b. "The Value of Trade and Regional Investments in the Pan-Arab Electricity Market: Integrating Power Systems and Building Economies." Energy Sector Management Assistance Program (ESMAP) paper, Report 166033, World Bank.

World Bank. 2022a. "Egypt—Country Climate and Development Report." Report 177494, World Bank.

World Bank. 2022b. "The Employment Benefits of an Energy Transition in Egypt." Energy Sector Management Assistance Program (ESMAP) paper, Report 178399, World Bank.

World Bank. 2022c. "The Employment Benefits of an Energy Transition in Morocco." Energy Sector Management Assistance Program (ESMAP) paper, Report 178398, World Bank.

World Bank. 2022d. "Green Fiscal Reforms: Part Two of Strengthening Inclusion and Facilitating the Green Transition." EU Regular Economic Report 7, World Bank.

World Bank. 2022e. "Guidelines on Integrating Nature-Based Passive Cooling Options into Urban Planning and Design." Energy Sector Management Assistance Program (ESMAP) report, World Bank.

World Bank. 2022f. "Piloting Nature-Based Solutions for Urban Cooling: Overview." Energy Sector Management Assistance Program (ESMAP) report, World Bank.

World Bank. 2022g. *Unlocking Sustainable Private Sector Growth in the Middle East and North Africa: Evidence from the Enterprise Survey.* European Investment Bank, European Bank for Reconstruction and Development, and World Bank.

World Bank. 2023a. *The Business of the State.* World Bank. doi:10.1596/978-1-4648-1998-8.

World Bank. 2023b. "Zambia—Second Macroeconomic Stability, Growth, and Competitiveness Development Policy Financing." Program document, Report PGD461, World Bank.

World Bank. 2023c. *Reality Check: Lessons from 25 Policies Advancing a Low-Carbon Future.* Climate Change and Development Series. World Bank. http://hdl.handle.net/10986/40262.

World Bank. 2024a. *Finance and Prosperity 2024. Special Focus: Sovereign-Bank Nexus Climate and the Banking Sector.* World Bank.

World Bank. 2024b. "Pakistan Blue Economy: Assessment of Monitoring of Risks to Sustainability and Grounding Evidence for Decision-Making." Unpublished Technical Note, World Bank.

World Bank. 2025a. "Business Ready 2025." Report 207546, World Bank.

World Bank. 2025b. "Climate and Development Brief: Scaling Climate Finance to Lower Emissions in the Power Sector." Report 202479, World Bank.

World Bank. 2025c. "Firms and Climate Adaptation Policy Toolkit Guidance Note." Unpublished Technical Note, World Bank.

World Bank. 2025d. "Jobs in a Changing Climate: Insights from World Bank Group Country Climate and Development Reports Covering 93 Economies." Report 206431, World Bank.

World Bank. 2025e. "Unlocking Efficiency: The Global Landscape of Building Energy Regulations and Their Enforcement." Report 200405, World Bank.

Zaveri, E. D., R. Gatti, and A. Islam. 2024. "Extreme Weather Shocks and Firms in the Middle East and North Africa." Policy Research Working Paper 11004, World Bank.

Zerbib, O. D. 2019. "The Effect of Pro-Environmental Preferences on Bond Prices: Evidence from Green Bonds." *Journal of Banking and Finance* 98: 39–60.

Appendix A: Annexes for Chapter 1

ANNEX 1A. THE PRIVATE SECTOR LANDSCAPE: SUPPLEMENTARY FIGURES

FIGURE 1A.1

Significant Business Environment Obstacles to Firms in MENAAP Countries

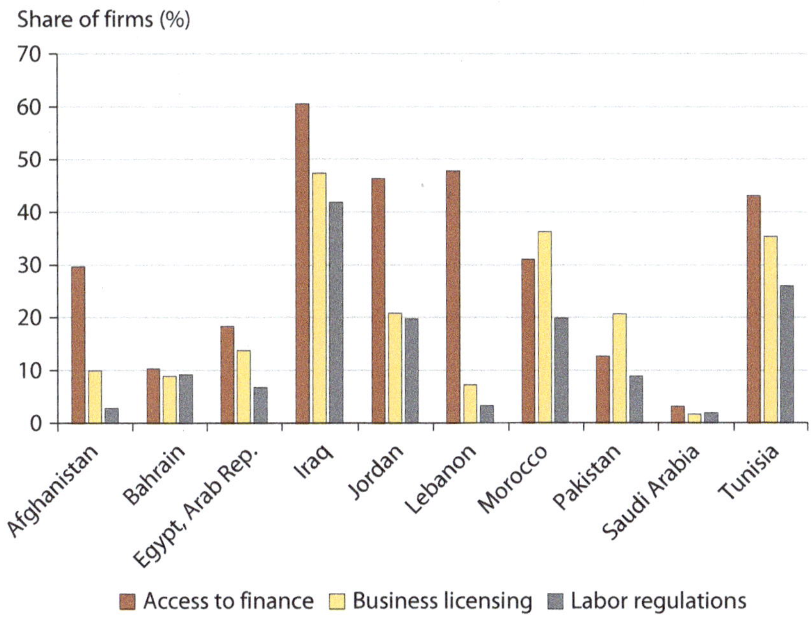

Source: Calculations based on World Bank Enterprise Surveys (WBES) data.
Note: An issue is counted as an obstacle if the firm reports it to be a major or very severe constraint. Summary statistics are computed using sampling weights. The sample includes a total of over 10,000 firms surveyed in Afghanistan (2025), Bahrain (2024), Egypt (2020), Iraq (2022), Jordan (2024), Lebanon (2019), Morocco (2023), Pakistan (2022), Saudi Arabia (2022), and Tunisia (2024). MENAAP = Middle East, North Africa, Afghanistan, and Pakistan.

FIGURE 1A.2

Most Firms in the MENAAP Region Remain Small

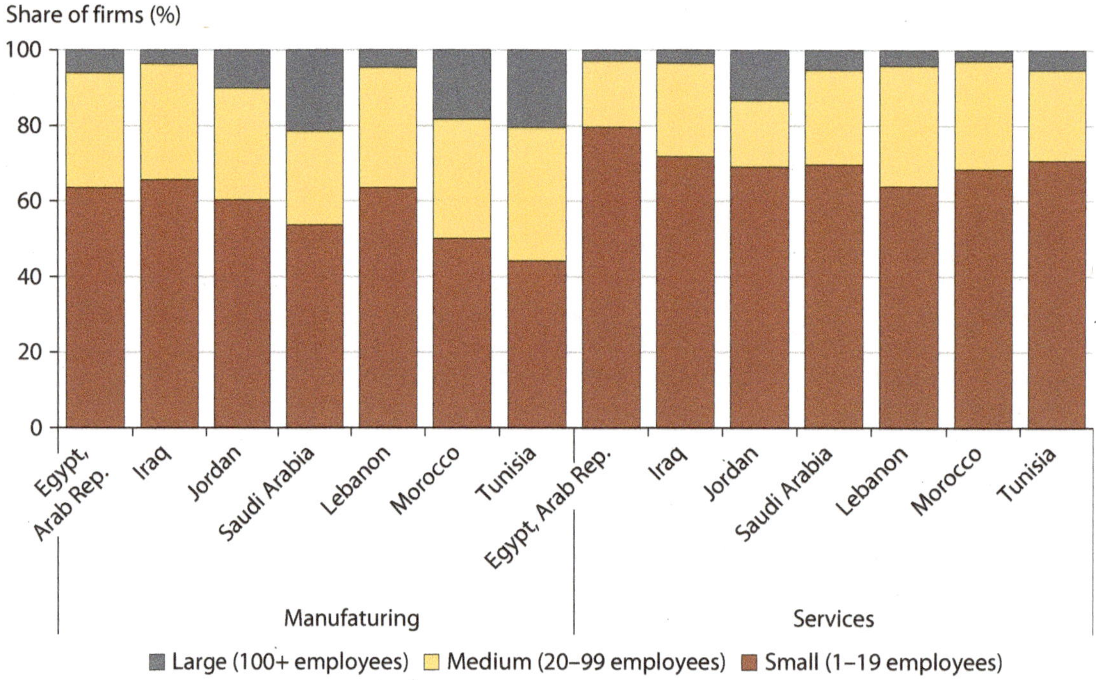

Source: Calculations based on World Bank Enterprise Surveys (WBES) data.

Note: Summary statistics computed using sampling weights. The sample includes a total of over 10,000 firms surveyed in Egypt, Arab Rep. (2020); Iraq (2022); Jordan (2024); Lebanon (2019); Morocco (2023); Saudi Arabia (2022); and Tunisia (2024). MENAAP = Middle East, North Africa, Afghanistan, and Pakistan.

FIGURE 1A.3

Most MENAAP Firms Are Either Not Involved in International Trade or Only Import

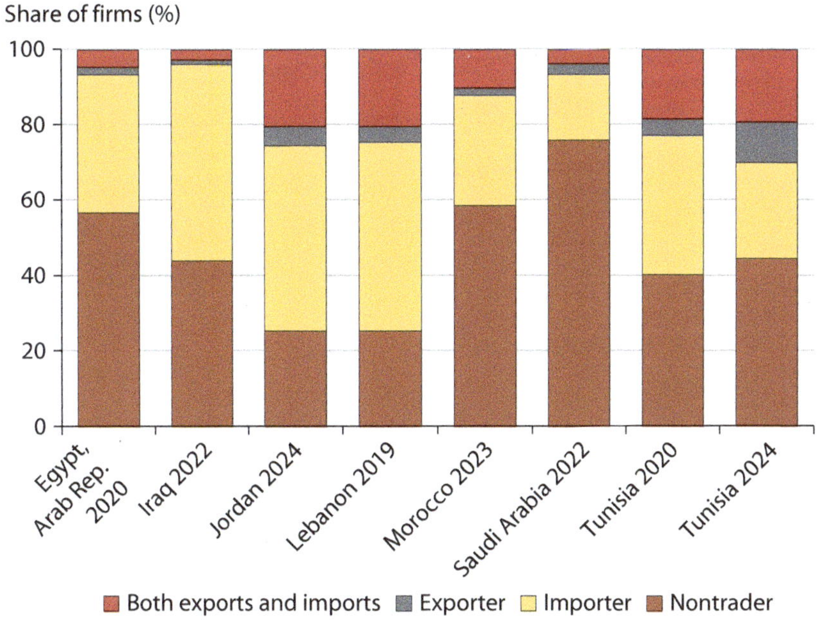

Source: Calculations based on World Bank Enterprise Surveys (WBES) data.
Note: Summary statistics computed using sampling weights. The WBES sample includes a total of over 10,000 firms surveyed in Egypt, Arab Rep. (2020); Iraq (2022); Jordan (2024); Lebanon (2019); Morocco (2023); Saudi Arabia (2022); and Tunisia (2020 and 2024). MENAAP = Middle East, North Africa, Afghanistan, and Pakistan.

ANNEX 1B. CONCEPTUAL FRAMEWORK FOR ANALYSIS OF CLIMATE EFFECTS ON FIRM PERFORMANCE AND RESULTING POLICIES

Direct and Indirect Channels of Negative Effects

The theoretical framework for firms and climate analysis distinguishes between direct (internal to the firm) and indirect (external to the firm) impacts of hazards with a slow onset, such as rising temperatures (World Bank 2025), as shown in figure 1B.1.

FIGURE 1B.1

Channels of Effects on Firms of Climate Hazards with a Slow Onset

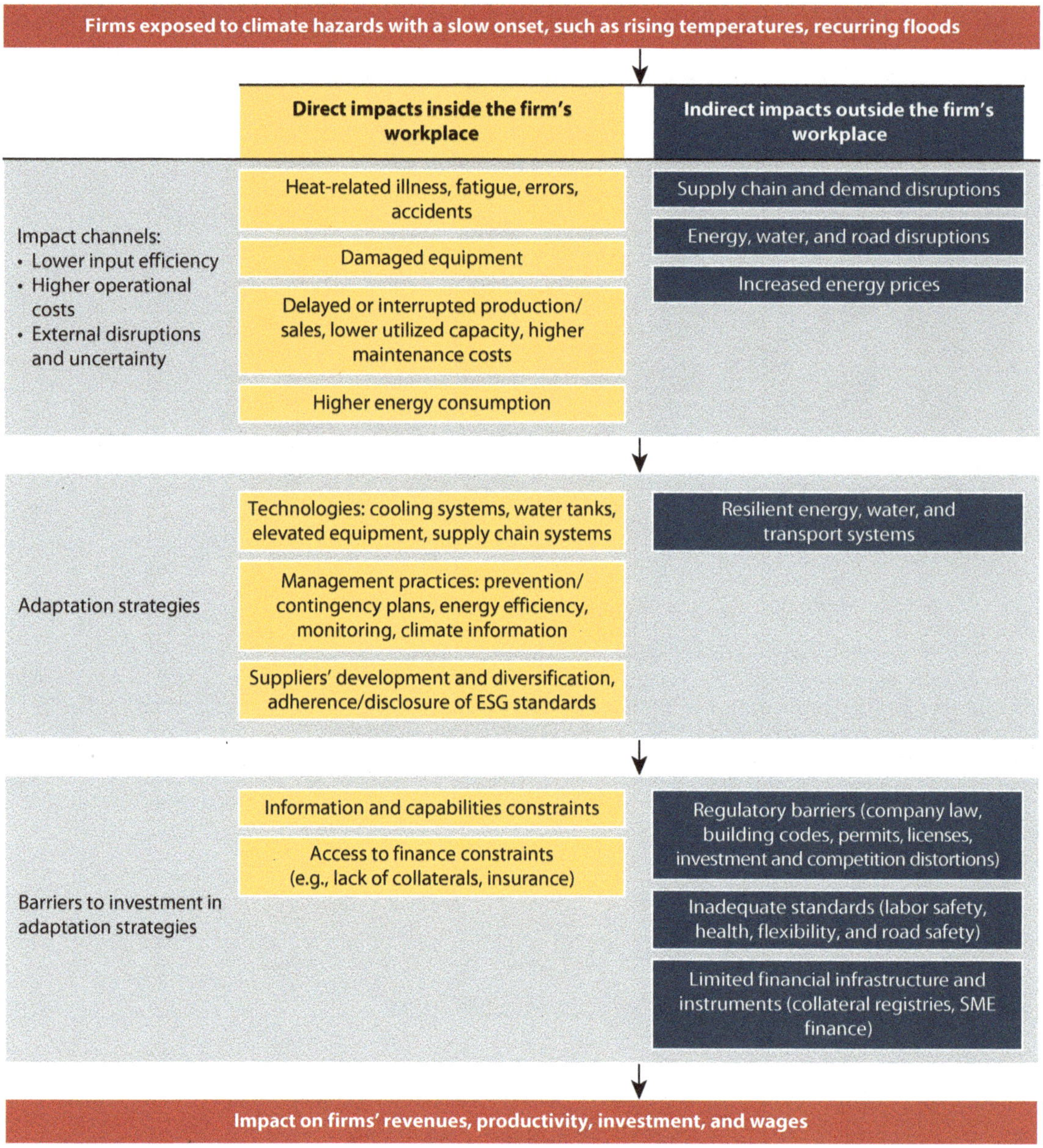

Source: World Bank Firms and Climate Adaptation Policy Toolkit (World Bank 2025).
Note: Examples of hazards with a slow onset include rising temperatures and recurring floods from precipitation. Yellow boxes are direct impacts inside the firm's workplace; light blue boxes are impacts that can occur either inside or outside the firm's workplace; dark blue boxes are indirect impacts outside the firm's workplace. ESG = environmental, social, and governance.

Prioritizing Policy Areas Based on Complementarity Between Adaptation Investments and Productivity

Governments can prioritize policy areas for climate resilience by assessing firms' vulnerabilities and understanding how adaptation investments relate to private sector growth. A key organizing principle is the degree of complementarity between adaptation investments and firm productivity (World Bank 2025). This can be assessed empirically by regressing adaptation investment on productivity and vice versa. Although this approach offers valuable insights, interpreting the strength of complementarity involves some judgment. A strong positive relationship suggests high complementarity, a flat or negative relationship indicates a trade-off, and moderate cases are those where the relationship is positive but adaptation remains somewhat limited even among productive firms. This complementarity allows governments to prioritize policy areas (refer to figure 1B.2).

When the complementarity is strong—firms that invest in adaptation also experience productivity gains—governments can consider prioritizing low-cost awareness programs. These programs, which highlight the adverse effects of climate stress, are typically low in cost and can motivate firms to invest when the incentive is already clear. In such settings, business environment reforms (such as improving access to finance or simplifying licensing processes) can further encourage private adaptation by reducing barriers to investment.

In cases of moderate complementarity, where the relationship between adaptation and productivity is positive but relatively weak—such as when only a minority of highly productive firms adopt adaptation practices—policy responses need to be more carefully targeted. A mix of instruments is likely required, but their design should reflect the uneven incentives firms face. For instance, business environment reforms may be most effective when focused on firms or regions that are especially vulnerable to climate risks rather than being applied broadly. In these settings, public infrastructure investments can play a supportive role by lowering the cost or increasing the feasibility of adaptation for firms that might otherwise hold back.

However, the effectiveness of these approaches depends heavily on country-specific conditions, such as market access, governance, or credit availability. In environments where these constraints are binding, the link between firm performance and adaptation may be weak, leaving firms with little incentive to invest on their own. In these cases, public infrastructure investments become an even more relevant policy tool—providing public goods that reduce exposure or facilitate adaptation directly, especially when private investment is unlikely to materialize without external support.

FIGURE 1B.2

The Complementarity Between Adaptation and Productivity Informs the Prioritization of Policy Areas

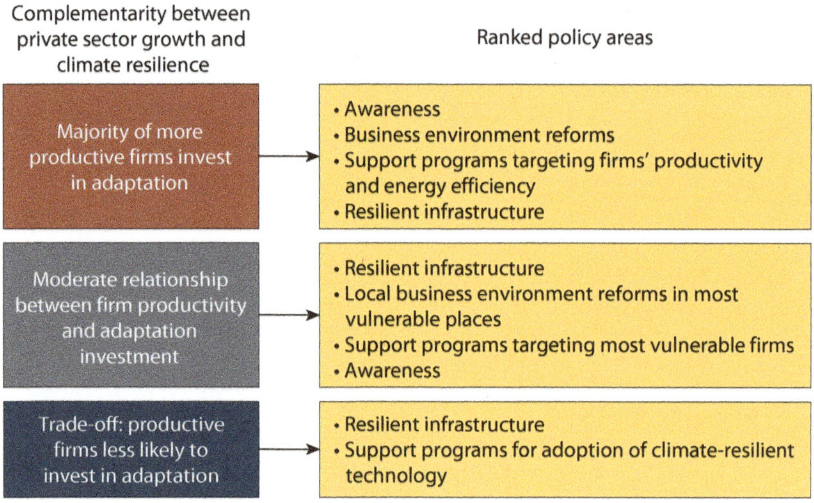

Source: World Bank Firms and Climate Adaptation Policy Toolkit (World Bank 2025).
Note: "Business environment" is broadly defined as the collection of laws, regulations, policies, and institutions that shape and drive business activity. It therefore includes foundational policy aspects such as business regulation, competition policy, and investment policy as well as policies enabling the environment of financial institutions.

ANNEX 1C. HAZARDS AND EXPOSURE: SUPPLEMENTARY FIGURES

FIGURE 1C.1

Average Temperatures Are Rising (1950–2023) in MENAAP Countries

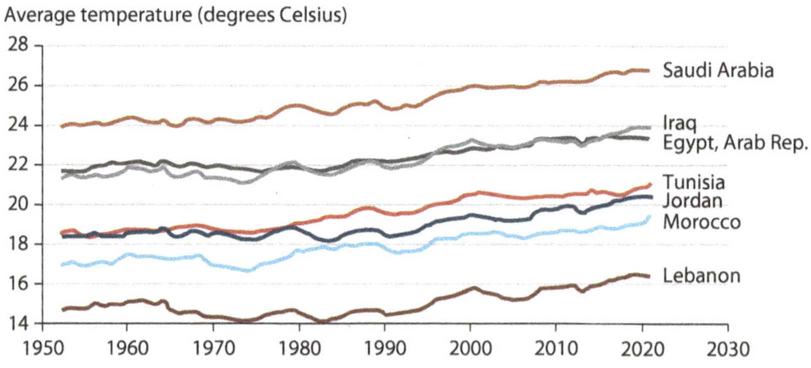

Source: Developed using the World Bank Firms and Jobs Vulnerability to Climate Change Explorer (https://datanalytics.worldbank.org/global-weather-map/).
Note: Temperature is defined as the temperature of air at two meters above the surface of land. The data are at the hourly mean level, which is then used to calculate a location's daily average temperature. MENAAP = Middle East, North Africa, Afghanistan, and Pakistan.

FIGURE 1C.2

Average Increases in Temperatures Varied Broadly Across MENAAP Countries Between 1990–95 and 2018–23

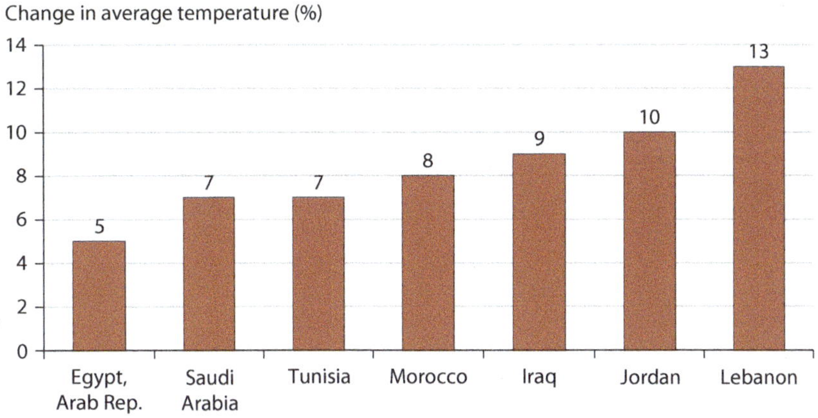

Source: World Bank Firms and Climate Adaptation Policy Toolkit (World Bank 2025).
Note: Temperature is defined as the temperature of air at two meters above the surface of land. The data are at the hourly mean level, which is then used to calculate a location's daily and yearly average temperature. The average yearly temperature is taken for the periods 1990–95 and 2018–23 for each country. The chart shows the change between these period means. MENAAP = Middle East, North Africa, Afghanistan, and Pakistan.

FIGURE 1C.3

The Average Number of Days with Extreme Heat Has Increased Across MENAAP Countries Between 1990–95 and 2018–23

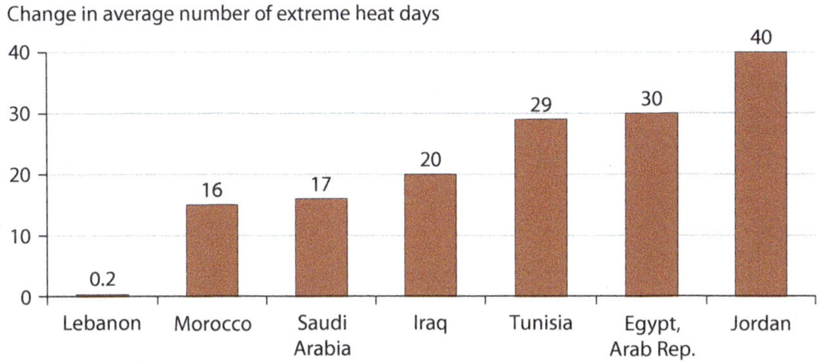

Source: World Bank Firms and Climate Adaptation Policy Toolkit (World Bank 2025).
Note: Extreme heat days are those whose average temperature exceeds 35 degrees Celsius. The average yearly number of extreme heat days is found for the periods 1990–95 and 2018–23 for each country. The chart shows the change between these period means. MENAAP = Middle East, North Africa, Afghanistan, and Pakistan.

MAP 1C.1

There Is Wide Variation in Exposure to Extreme Heat Days Across and Within MENAAP Countries, 2023

a. Morocco

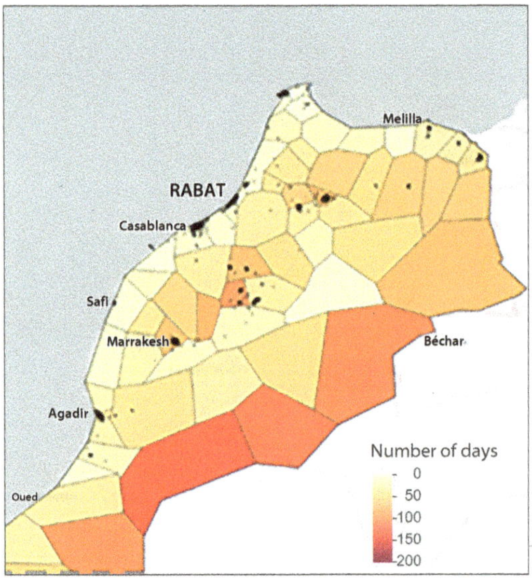

b. Tunisia

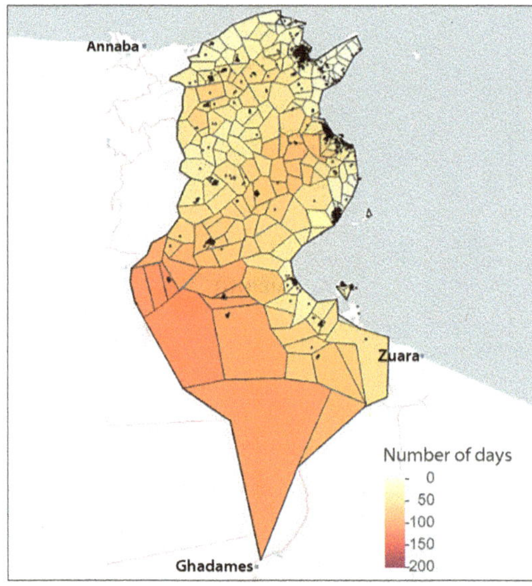

c. Egypt, Arab Rep.

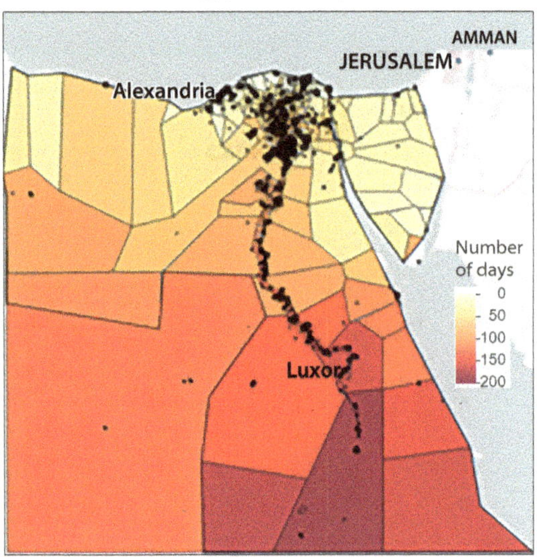

d. Iraq

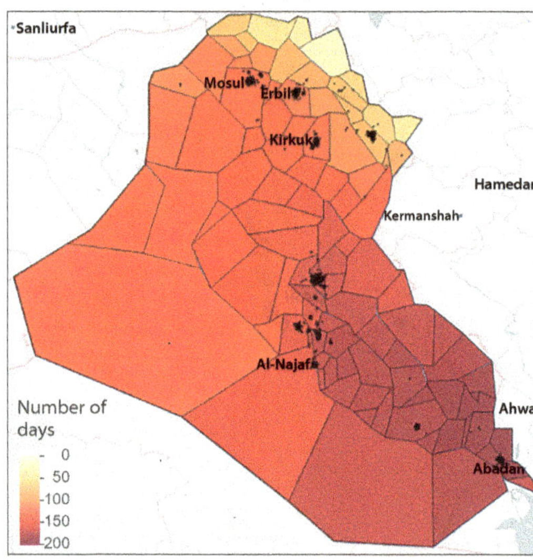

(continued)

MAP 1C.1

There is Wide Variation in Exposure to Extreme Heat Days Across and within MENAAP Countries, 2023 (*continued*)

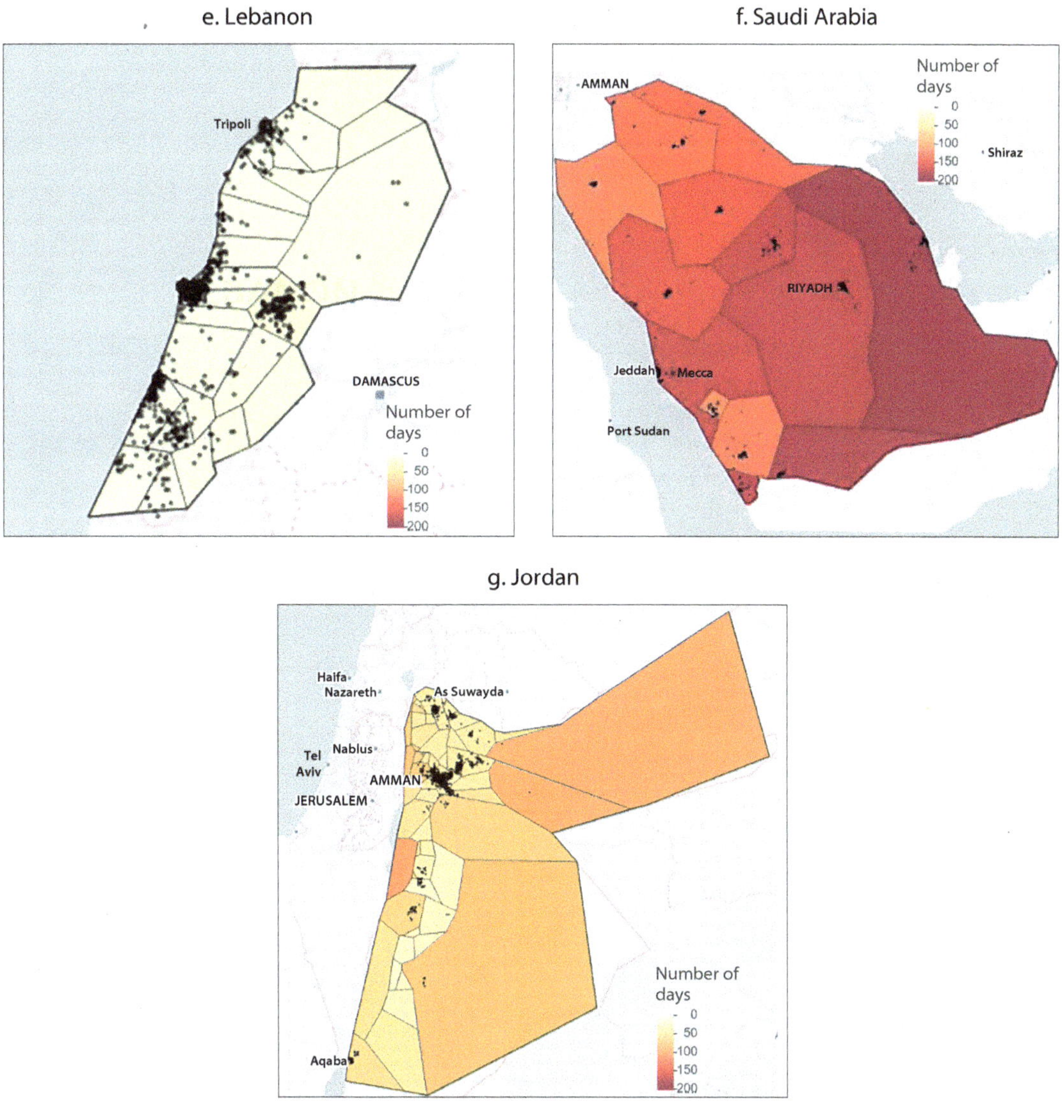

Source: Developed using the World Bank Firms and Jobs Vulnerability to Climate Change Explorer (https://datanalytics.worldbank.org/global-weather-map/).
Note: The heat maps show the number of extreme heat days per district (days whose maximum temperature exceeds 35 degrees Celsius) in 2023. The color coding is based on worldwide geocoded daily satellite weather data in 2023. Black dots are locations of manufacturing and services firms from the World Bank Enterprise Surveys (WBES). The toolkit's dataset combines geocoded firm-level data from WBES with weather data from the Database of Regional Climate Change (DRCC) developed by Bettarelli et al. 2024. It allows for the identification of key climate change risks (hazards) and the clusters of firms they might affect (exposure). The evolution of climate risks from heat, droughts, and floods can be measured and compared over a period spanning from 1950 to 2023. MENAAP = Middle East, North Africa, Afghanistan, and Pakistan.

ANNEX 1D. VARIABLE DESCRIPTIONS AND REGRESSION SPECIFICATIONS FOR DATA ANALYSIS ON THE EFFECTS OF CLIMATE CHANGE ON FIRMS

Firm Performance Indicators

- *Sales (d2):* total sales in last fiscal year
- *Employment (l1):* number of permanent, full-time employees at end of last fiscal year
- *Productivity (d2/l1):* sales per worker
- *Wages (n2a):* total labor cost in last fiscal year
- *Value added (d2 – n2e):* sales minus cost of raw materials and intermediate goods used in production in last fiscal year
- *Investment in fixed assets (k4):* dummy variable taking value of 1 if establishment purchased any fixed assets in last fiscal year
- *Capacity utilization (f1):* output produced as a percentage of the maximum possible output given the available resources in last fiscal year
- *Insufficient water supply (c15):* dummy variable taking value of 1 if establishment reports insufficient water supply as being a problem it faced in last fiscal year
- *Electricity cost (n2b):* total annual cost of electricity in last fiscal year
- *Energy intensity (n2b/d2):* electricity cost over sales
- *Power outages (c6):* dummy variable taking value of 1 if establishment experienced a power outage in the last fiscal year
- *Electricity consumption (BMGE.1):* in fiscal year [insert last complete fiscal year], how much electricity, in kilowatt-hours, did this establishment consume?
- *Transport obstacle (d30a):* dummy variable taking value of 1 if establishment reports transport as a moderate, major, or very severe obstacle

Climate Variables

- *Heat days deviation from long-run trend:* average number of local extreme heat days in last fiscal year minus local long-run trend (calculated between 1980 and 2008). An extreme heat day is defined as a day when the maximum daily temperature anywhere in the district (admin-2 geographical granularity) exceeds 35 degrees Celsius. Each firm falls into a district and takes on that district's number of extreme heat days.
- *Average temperature deviation from long-run trend:* average temperature (Celsius) in last fiscal year minus local long-run trend (calculated between 1980 and 2008).

Firm Characteristics and Policy Variables

- *Regulation index:* composed of "Business licensing is an obstacle" (j30c), "Labor regulation is an obstacle" (l30a), and "Firm inspected by tax officials" (j3), coded such that higher values represent greater regulatory burdens
- *Access to finance index:* composed of "Firm has checking or savings account" (k6), "Firm has overdraft facility" (k7), and "Firm has credit or loan from a financial institution" (k8), coded such that higher values represent reduced access to finance
- *Labor force index:* composed of "Inadequately educated workforce is an obstacle" (l30b) and "formal training programs for employees" (l10), coded such that higher values represent a less educated or trained workforce
- *Sector:* manufacturing or services as defined in the WBES
- *Small:* dummy variable taking value of 1 if firm has between 5 and 19 employees
- *Young:* dummy variable taking value of 1 if firm is 5 years old or younger
- *GVC participant:* dummy variable taking value of 1 if firm does any of the following: imports (uses foreign inputs), exports (a positive share of its sales is sold abroad), uses foreign technology, and/or is foreign owned (more than 10 percent of ownership is foreign)

- *Foreign owned:* dummy variable taking value of 1 if more than 10 percent of the firm's ownership is foreign
- *Part of larger firm:* dummy variable taking value of 1 if the firm is part of a multiestablishment firm

Regression Specifications

- Direct effects: $Y_{isct} = \mu_{cs} + \lambda_t + \beta CCS_{it} + \varepsilon_{it}$ (1D.1)
  - Regress firm performance indicators (*Y*) on country, sector, country-sector (μ) and year (λ) fixed-effects and the relevant climate change shock (*CCS*). Firm-level size and age controls are also included.
- Interactions: $Y_{isct} = \mu_{cs} + \lambda_t + \beta CCS_{it} * Characteristic/Policy_{it} + \varepsilon_{it}$ (1D.2)
  - Regress firm performance indicators (*Y*) on country, sector, country-sector (μ) and year (λ) fixed-effects and the relevant climate change shock (*CCS*) interacted with a firm characteristic or a policy variable (for example, business environment quality index). Firm-level size and age controls are also included.

ANNEX 1E. VULNERABILITY RESULTS

Heat-vulnerable firms are more likely to invest in fixed assets as well as adaptation practices. As temperatures may drive up repair and maintenance needs, firms might invest in replacing or upgrading physical assets. There is also positive correlation between a firm's investment in fixed assets and the likelihood that it invests in cooling systems, monitors water usage, and tracks emissions. Similar results are found on the effects of average temperatures on input efficiency and operating costs.
The regressions shown in this annex illustrates those correlations.

Effects of Heat Days on Firm Performance (Input Efficiency Channel)

TABLE 1E.1

Effects of Extreme Heat Days on Main Firm Performance Outcomes in MENAAP Countries

Dependent variable	Sales (log)	Sales per worker (log)	Workers (log)	Wages (log)	Capital utilization (log)	Investment in fixed assets (dummy)	Investment in fixed assets (no Iraq)
Heat days (dev LT trend)	−0.061*** (0.023)	−0.038* (0.018)	−0.022* (0.013)	−0.080*** (0.019)	−0.030*** (0.007)	0.019*** (0.005)	0.005 (0.005)
Obs	9,416	9,390	10,065	9,316	4,578	10,002	9,002
R2	0.665	0.723	0.591	0.724	0.132	0.105	0.110

Source: Table original to this publication. Calculations based on the World Bank Firms and Climate Adaptation Policy Toolkit (World Bank 2025) using World Bank Enterprise Surveys (WBES) and Database of Regional Climate Change data (Bettarelli et al. 2024).
Note: "Extreme heat" days are those whose maximum temperature exceeds 35 degrees Celsius. Geocoded satellite weather data were matched with geocoded WBES data for over 10,000 firms across Egypt, Arab Rep.; Iraq; Jordan; Lebanon; Morocco; Saudi Arabia; and Tunisia between 2019 and 2024. All regressions include controls for long-term (LT) means of heat days, firm size, and age, as well as country, sector, year, and country-sector fixed effects (FE). Regressions are not weighted. Results are robust to removing country-sector FE and/or firm size and age controls. Removing Iraq from the sample renders the effect of heat days on investment insignificant. Effects of heat days on the following are not significant: value added, value added per worker. MENAAP = Middle East, North Africa, Afghanistan, and Pakistan.
Significance level: * = 10 percent, *** = 1 percent.

TABLE 1E.2

Effects of Extreme Heat Days Interacted with Sectors on Sales, MENAAP Countries

Dependent variable	Sales (log)				
Base heat days (dev LT trend) / effect for all other sectors	−0.068*** (0.023)	−0.068*** (0.024)	−0.056** (0.025)	−0.056** (0.024)	−0.060** (0.025)
Interaction (heat days*sector)	Textiles/ garments 0.107* (0.061)	Food 0.043 (0.042)	Other manufacturing −0.028 (0.036)	Retail/ wholesale −0.049 (0.040)	Other services −0.031 (0.037)
Combined effect for sector	0.039	−0.025	−0.084	−0.105	−0.091
Obs	9,416	9,416	9,416	9,416	9,416
R2	0.664	0.663	0.664	0.663	0.663
Country and year FE	Yes	Yes	Yes	Yes	Yes
Country-sector FE	No	No	No	No	No
Controls	Yes	Yes	Yes	Yes	Yes

Source: Table original to this publication. Calculations based on the World Bank Firms and Climate Adaptation Policy Toolkit (World Bank 2025) using World Bank Enterprise Surveys (WBES) and Database of Regional Climate Change data (Bettarelli et al. 2024).
Note: "Extreme heat" days are those whose maximum temperature exceeds 35 degrees Celsius. Geocoded satellite weather data were matched with geocoded WBES data for over 10,000 firms across Egypt, Arab Rep.; Iraq; Jordan; Lebanon; Morocco; Saudi Arabia; and Tunisia between 2019 and 2024. All regressions include controls for long-term (LT) means of heat days, country and year fixed effects, and firm size and age controls. Regressions are not weighted. FE = fixed effects; MENAAP = Middle East, North Africa, Afghanistan, and Pakistan.
Significance level: * = 10 percent, ** = 5 percent, *** = 1 percent.

Effects of Heat Days on Firm Performance (Operational Costs Channel)

TABLE 1E.3

Effects of Extreme Heat Days on Costs, MENAAP Countries

Dependent variable	Electricity cost (log)	Electricity cost (no Egypt)	Electricity cost (no Saudi Arabia)	Electricity cost (nonlinearity)	Insufficient water supply (dummy)	Power outages (dummy)
Heat days (dev LT trend)	−0.078*** (0.024)	−0.013 (0.038)	−0.019 (0.026)	Linear term: 0.150*** Squared term: −0.083***	0.019*** (0.006)	−0.003 (0.006)
Obs	9,249	6,196	7,844	9,249	4,909	10,059
R2	0.678	0.754	0.701	0.678	0.096	0.103

Source: Table original to this publication. Calculations based on the World Bank Firms and Climate Adaptation Policy Toolkit (World Bank 2025) using World Bank Enterprise Surveys (WBES) and Database of Regional Climate Change data (Bettarelli et al. 2024).
Note: "Extreme heat" days are those whose maximum temperature exceeds 35 degrees Celsius. Geocoded satellite weather data were matched with geocoded WBES data for over 10,000 firms across Egypt, Arab Rep.; Iraq; Jordan; Lebanon; Morocco; Saudi Arabia; and Tunisia between 2019 and 2024. All regressions include controls for long-term (LT) means of heat days, country, sector, year, and country-sector fixed effects (FE), and firm size and age controls. Regressions are not weighted. Results are robust to removing country-sector FE and/or firm size and age controls. Removing either Egypt or Saudi Arabia from the sample renders the effect of heat days on electricity cost insignificant (results possibly influenced by broad energy tariff structures in both countries and by Saudi Arabia heavily subsidizing energy consumption). Effects of heat days on the following are not significant: power outages. MENAAP = Middle East, North Africa, Afghanistan, and Pakistan.
Significance level: *** = 1 percent.

Heterogeneity of Impacts (Interactions of Heat Days with Firms' Characteristics)

TABLE 1E.4

Effects of Extreme Heat Days Interacted with Firms' Characteristics on Sales, MENAAP Countries

Dependent variable	Sales (log)				
Heat days (dev LT trend)	−0.018 (0.031)	−0.070** (0.023)	−0.134** (0.026)	−0.078*** (0.023)	−0.080*** (0.023)
Interaction of heat days with…	Small −0.085*** (0.033)	Young 0.092* (0.051)	GVC 0.165*** (0.032)	Foreign owned 0.044 (0.072)	Part of larger firm 0.228*** (0.063)
Obs	9,416	9,416	9,416	9,211	9,409
R2	0.666	0.665	0.681	0.673	0.669

Source: Table original to this publication. Calculations based on the World Bank Firms and Climate Adaptation Policy Toolkit (World Bank 2025) using World Bank Enterprise Surveys (WBES) and Database of Regional Climate Change data (Bettarelli et al. 2024).
Note: "Extreme heat" days are those whose maximum temperature exceeds 35 degrees Celsius. Geocoded satellite weather data were matched with geocoded WBES data for over 10,000 firms across Egypt, Arab Rep.; Iraq; Jordan; Lebanon; Morocco; Saudi Arabia; and Tunisia between 2019 and 2024. All regressions include controls for long-term (LT) means of heat days, country, sector, year, and country-sector fixed effects (FE), and firm size and age controls. Regressions are not weighted. Results are mostly robust to removing country-sector FE and/or firm size and age controls. Small firms have 1–19 employees. Young firms are less than five years old. GVC = global value chain; MENAAP = Middle East, North Africa, Afghanistan, and Pakistan.
Significance level: * = 10 percent, ** = 5 percent, *** = 1 percent.

TABLE 1E.5

Effects of Extreme Heat Days Interacted with Firms' Characteristics on Sales per Worker, MENAAP Countries

Dependent variable	Sales per worker (log)				
Heat days (dev LT trend)	0.020 (0.025)	−0.045** (0.019)	−0.136*** (0.023)	−0.055*** (0.019)	−0.052*** (0.019)
Interaction	Small −0.115*** (0.028)	Young 0.072 (0.044)	GVC 0.206*** (0.028)	Foreign owned 0.091* (0.060)	Part of larger firm 0.152*** (0.051)
Obs	9,390	9,390	9,390	9,192	9,383
R2	0.724	0.723	0.730	0.728	0.724

Source: Table original to this publication. Calculations based on the World Bank Firms and Climate Adaptation Policy Toolkit (World Bank 2025) using World Bank Enterprise Surveys (WBES) and Database of Regional Climate Change data (Bettarelli et al. 2024).
Note: "Extreme heat" days are those whose maximum average temperature exceeds 35 degrees Celsius. Geocoded satellite weather data were matched with geocoded WBES data for over 10,000 firms across Egypt, Arab Rep.; Iraq; Jordan; Lebanon; Morocco; Saudi Arabia; and Tunisia between 2019 and 2024. All regressions include controls for long-term (LT) means of heat days, country, sector, year, and country-sector fixed effects (FE), and firm size and age controls. Regressions are not weighted. Results are robust to removing country-sector FE and/or firm size and age controls. Small firms have 1 to 19 employees. Young firms are less than five years old. GVC = global value chain. MENAAP = Middle East, North Africa, Afghanistan, and Pakistan. Significance level: * = 10 percent, ** = 5 percent, *** = 1 percent.

Effects of Average Temperatures on Firms' Performance

Rising average temperatures produce similar negative effects on firms' performance. A one standard deviation increase in average temperature (equivalent to 0.83 degrees Celsius) is associated with employment losses of 8 percent, a 6 percent decrease in wages, and an 8 percent drop in capital utilization (refer to table 1E.6). Additionally, the likelihood of experiencing water shortages increases by 5 percent.

Similar to the pattern observed for extreme heat days, firms appear more likely to invest in fixed assets as average temperatures rise. Allowing for nonlinear effects of climate by including a squared term in regressions reveals that investment declines at lower temperatures but increases at higher ones—possibly reflecting rising repairs and maintenance costs as heat exposure worsens.

Somewhat surprisingly, sales per worker increase with rising temperatures. This is likely a mechanical effect driven by falling employment while sales remain stable. The pattern is consistent with capital-labor substitution, whereby firms respond to heat stress by investing in capital while scaling back labor. Supporting evidence includes declining capital utilization, rising investment, and the finding that electricity costs are affected by extreme heat days but not by average temperatures (refer to table 1E.7). This suggests that heat waves disrupt both labor and capital via energy costs, while moderate temperature increases primarily affect labor but can be partially offset through substitution toward capital.

TABLE 1E.6

Effects of Average Temperature on Main Firm Performance Outcomes, MENAAP Countries

Dependent variable	Sales per worker (log)	Workers (log)	Wages (log)	Capital utilization (log)	Investment in fixed assets (dummy)	Investment in fixed assets (nonlinearity)
Temperature (dev LT trend)	0.118*** (0.027)	−0.081*** (0.017)	−0.063* (0.029)	−0.084*** (0.012)	0.049*** (0.007)	Linear term: −0.130*** (0.026) Squared term: 0.045*** (0.006)
Obs	9,390	10,065	9,361	4,578	10,002	10,002
R2	0.724	0.592	0.724	0.133	0.107	0.111

Source: Table original to this publication. Calculations based on the World Bank Firms and Climate Adaptation Policy Toolkit (World Bank 2025) using World Bank Enterprise Surveys (WBES) and Database of Regional Climate Change data (Bettarelli et al. 2024).
Note: Geocoded satellite weather data were matched with geocoded WBES data for over 10,000 firms across Egypt, Arab Rep.; Iraq;, Jordan; Lebanon; Morocco; Saudi Arabia; and Tunisia between 2019 and 2024. All regressions include controls for long-term (LT) means of heat days, country, sector, year, and country-sector fixed effects (FE), and firm size and age controls. Regressions are not weighted. Results are robust to removing country-sector FE and/or firm size and age controls. Effects of temperature on the following are not significant: sales, value added, value added per worker. MENAAP = Middle East, North Africa, Afghanistan, and Pakistan.
Significance level: * = 10 percent, *** = 1 percent.

TABLE 1E.7

Effects of Average Temperature on Firms' Costs, MENAAP Countries

Dependent variable	Insufficient water supply (dummy)	Power outages (dummy)
Temperature (dev LT trend)	0.046*** (0.009)	0.010 (0.008)
Obs	4,909	10,059
R2	0.103	0.104

Source: Table original to this publication. Calculations based on the World Bank Firms and Climate Adaptation Policy Toolkit (World Bank 2025) using World Bank Enterprise Surveys (WBES) and Database of Regional Climate Change data (Bettarelli et al. 2024).
Note: Geocoded satellite weather data were matched with geocoded WBES data for over 10,000 firms across Egypt, Arab Rep.; Iraq; Jordan; Lebanon; Morocco; Saudi Arabia; and Tunisia between 2019 and 2024. All regressions include controls for long-term (LT) means of heat days, country, sector, year, and country-sector fixed effects (FE), and firm size and age controls. Regressions are not weighted. Results are robust to removing country-sector FE and/or firm size and age controls. Effects of temperature on the following are not significant: electricity cost, power outages. MENAAP = Middle East, North Africa, Afghanistan, and Pakistan.
Significance level: *** = 1 percent.

ANNEX 1F. RESILIENCE RESULTS: SUPPLEMENTARY TABLES AND FIGURES

TABLE 1F.1

Effects of Heat Days Interacted with Adaptation Practices on Sales per Worker, MENAAP Countries

Dependent variable	Sales per worker (log)			
Heat days (dev LT trend)	−0.011 (0.028)	−0.015 (0.027)	−0.048 (0.029)	−0.039 (0.027)
Interaction	Heating/cooling 0.186*** (0.055)	Water management 0.143* (0.080)	Lighting systems 0.087* (0.045)	Climate-related strategic objectives 0.234*** (0.080)
Obs	4,332	4,374	4,657	4,980
R2	0.702	0.694	0.685	0.672

Source: Table original to this publication. Calculations based on the World Bank Firms and Climate Adaptation Policy Toolkit (World Bank 2025) using World Bank Enterprise Surveys (WBES), the Database of Regional Climate Change data (Bettarelli et al. 2024), and "green management scores" (De Haas et al. 2025).
Note: "Extreme heat" days are those whose maximum temperature exceeds 35 degrees Celsius. The WBES sample includes almost 6,000 firms across Egypt, Arab Rep.; Lebanon; Jordan; Morocco; and Tunisia between 2019 and 2020. All regressions include controls for long-term (LT) means of heat days, country, sector, year, and country-sector fixed effects, and firm size and age controls. Regressions are not weighted. Effects of heat days interacted with "green management scores" (De Haas et al. 2025) are not significant. MENAAP = Middle East, North Africa, Afghanistan, and Pakistan.
Significance level: * = 10 percent, *** = 1 percent.

TABLE 1F.2

Effects of Access to Finance on Adopting Cooling Improvements, MENAAP Countries

Dependent variable	Adopted heating/cooling improvements
Access to finance index coefficient	0.183*** (0.027)
Average marginal effect	0.049
Obs	4,685

Source: Table original to this publication. Calculations based on the World Bank Firms and Climate Adaptation Policy Toolkit (World Bank 2025) using World Bank Enterprise Surveys (WBES) and Database of Regional Climate Change data (Bettarelli et al. 2024).
Note: The WBES sample includes approximately 6,000 firms across Egypt, Arab Rep.; Jordan; Lebanon; Morocco; and Tunisia between 2019 and 2020 (smaller samples in regressions are due to nonresponses). All regressions include controls for the long-term mean of heat days, country and sector fixed effects, and firm size and age controls. Regressions are not weighted. Conditions improve as the index gets higher. Interpretation example (using average marginal effect): a one-unit increase in the access to finance index (meaning better access) increases the probability that a firm will adopt cooling technologies by 4.9 percentage points, holding all other variables constant. Effects of labor force index on adopting cooling are not significant. MENAAP = Middle East, North Africa, Afghanistan, and Pakistan.
Significance level: *** = 1 percent.

Figures 1F.1 through 1F.5 present the share of firms across MENAAP countries that have adopted water management practices, monitored water usage and emissions, improved lighting systems, and set climate-related strategic objectives, disaggregated by firm size. Figure 1F.6 illustrates the relationship between firm productivity and the likelihood of investing in heating or cooling improvements across MENAAP countries. Figure 1F.7 illustrates the relationship between firm productivity and the likelihood of monitoring water usage across MENAAP countries.

FIGURE 1F.1

Share of Firms that Adopted Water Management Practices, by Firm Size, MENAAP Countries

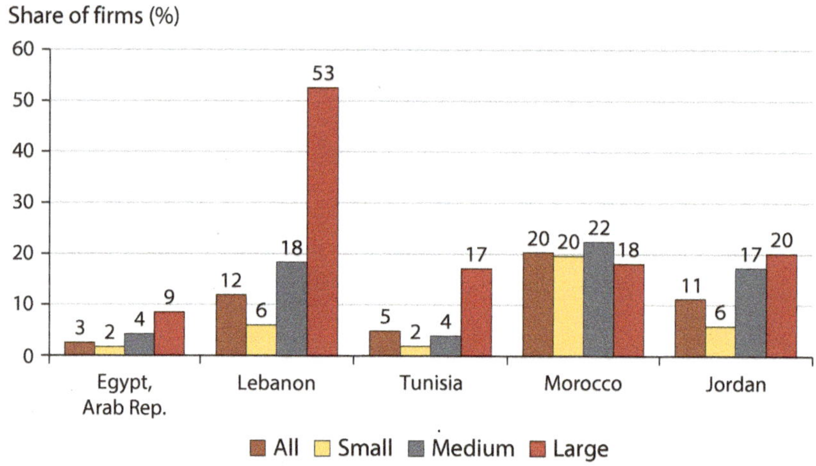

Source: World Bank Enterprise Surveys (WBES) data.
Note: The WBES sample includes almost 6,000 firms across Egypt, Arab Rep.; Jordan; Lebanon; Morocco; and Tunisia between 2019 and 2020. Summary statistics computed using sampling weights. Small firms are those with 5–19 employees; medium firms, 20–99 employees; and large firms, 100 or more employees. MENAAP = Middle East, North Africa, Afghanistan, and Pakistan.

FIGURE 1F.2

Share of Firms Monitoring Water Usage, by Firm Size, MENAAP Countries

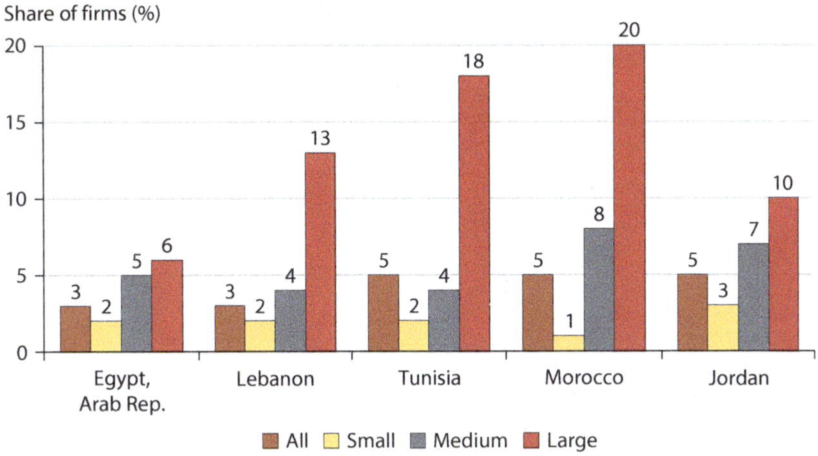

Source: World Bank Enterprise Surveys (WBES) data.
Note: The WBES sample includes almost 6,000 firms across Egypt, Arab Rep.; Jordan; Lebanon; Morocco; and Tunisia between 2019 and 2020. Summary statistics computed using sampling weights. Small firms are those with 5–19 employees; medium firms, 20–99 employees; and large firms, 100 or more employees. MENAAP = Middle East, North Africa, Afghanistan, and Pakistan.

FIGURE 1F.3

Share of Firms that Improved Lighting Systems, by Firm Size, in MENAAP Countries

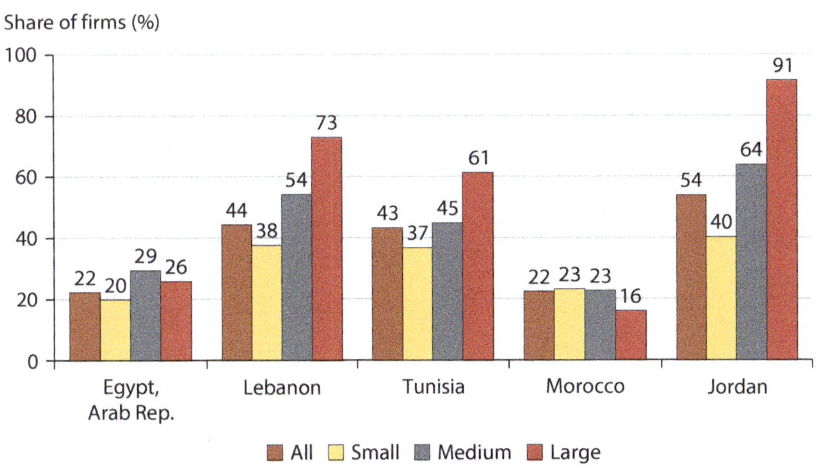

Source: World Bank Enterprise Surveys (WBES) data.
Note: The WBES sample includes almost 6,000 firms across Egypt, Arab Rep.; Jordan; Lebanon; Morocco; and Tunisia between 2019 and 2020. Summary statistics computed using sampling weights. Small firms are those with 5–19 employees; medium firms, 20–99 employees; and large firms, 100 or more employees. MENAAP = Middle East, North Africa, Afghanistan, and Pakistan.

FIGURE 1F.4

Share of Firms that Monitor Emissions, by Firm Size, in MENAAP Countries

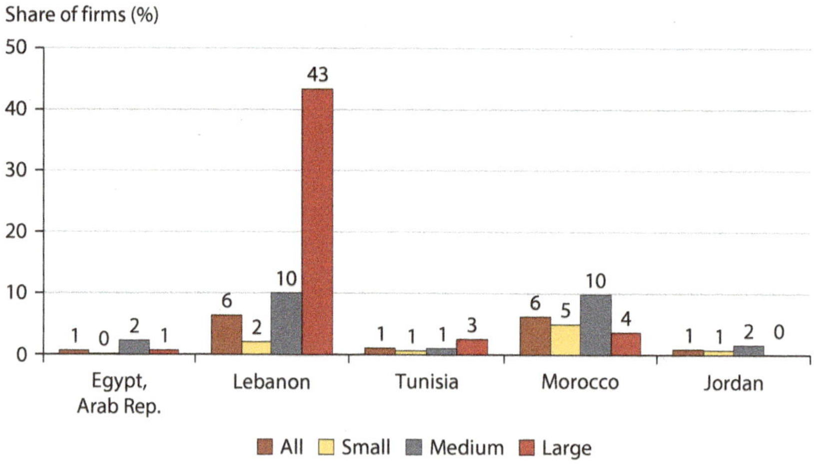

Source: World Bank Enterprise Surveys (WBES) data.
Note: The WBES sample includes almost 6,000 firms across Egypt, Arab Rep.; Jordan; Lebanon; Morocco; and Tunisia between 2019 and 2020. Summary statistics computed using sampling weights. Small firms are those with 5–19 employees; medium firms, 20–99 employees; and large firms, 100 or more employees. MENAAP = Middle East, North Africa, Afghanistan, and Pakistan.

FIGURE 1F.5

Share of Firms Setting Climate-Related Strategic Objectives, by Firm Size, MENAAP Countries

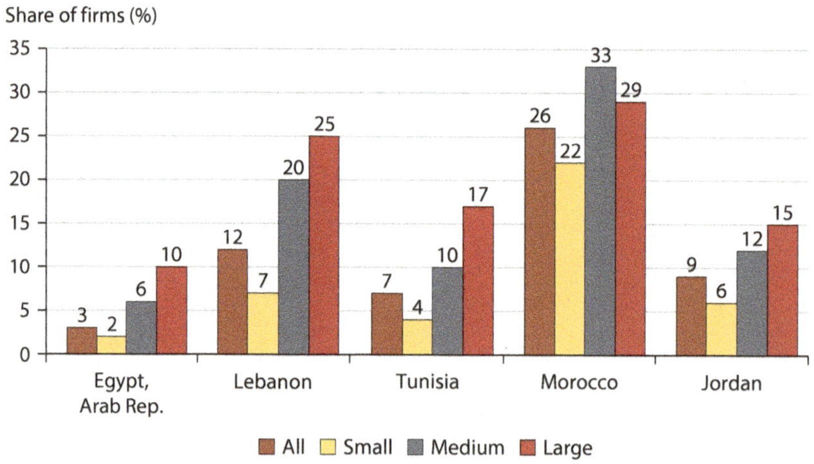

Source: World Bank Enterprise Surveys (WBES) data.
Note: The WBES sample includes almost 6,000 firms across Egypt, Arab Rep.; Jordan; Lebanon; Morocco; and Tunisia between 2019 and 2020. Summary statistics computed using sampling weights. Small firms are those with 5–19 employees; medium firms, 20–99 employees; and large firms, 100 or more employees. MENAAP = Middle East, North Africa, Afghanistan, and Pakistan.

FIGURE 1F.6

Complementary Between Firm Productivity and Probability of Investing in Heating or Cooling Improvements, MENAAP Countries

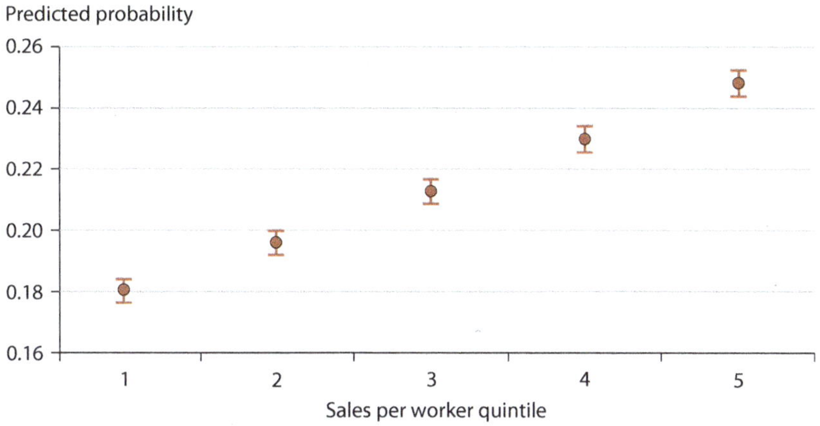

Source: Calculations based on the World Bank Firms and Climate Adaptation Policy Toolkit (World Bank 2025) using World Bank Enterprise Surveys (WBES) data.
Note: The WBES sample includes almost 6,000 firms across Egypt, Arab Rep.; Jordan; Lebanon; Morocco; and Tunisia between 2019 and 2020. Results based on probit regressions—dummy for adopting adaptation regressed on sales per worker quintile, country and sector fixed effects, and firm age and size controls. Each circle is the average predicted probability across all firms in the quintile. Error bars show the 95 percent confidence intervals. MENAAP = Middle East, North Africa, Afghanistan, and Pakistan.

FIGURE 1F.7

Complementarity Between Firm Productivity and Probability of Monitoring Water Usage, MENAAP Countries

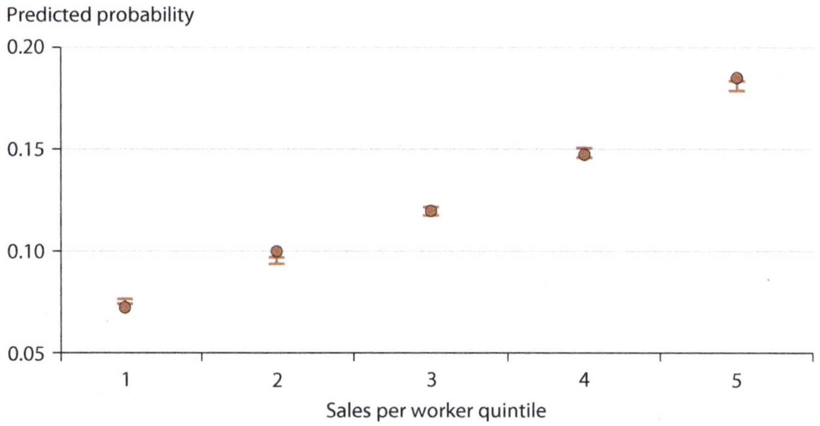

Source: Calculations based on the World Bank Firms and Climate Adaptation Policy Toolkit (World Bank 2025) using World Bank Enterprise Surveys (WBES) data.
Note: The WBES sample includes almost 6,000 firms across Egypt, Arab Rep.; Jordan; Lebanon; Morocco; and Tunisia between 2019 and 2020. Results based on probit regressions—dummy for adopting adaptation regressed on sales per worker quintile, country and sector fixed effects, and firm age and size controls. Each circle is the average predicted probability across all firms in the quintile. Error bars show the 95 percent confidence intervals. MENAAP = Middle East, North Africa, Afghanistan, and Pakistan.

FIGURE 1F.8

Complementarity Between Firm Productivity and Probability of Setting Climate-Related Strategic Objectives, MENAAP Countries

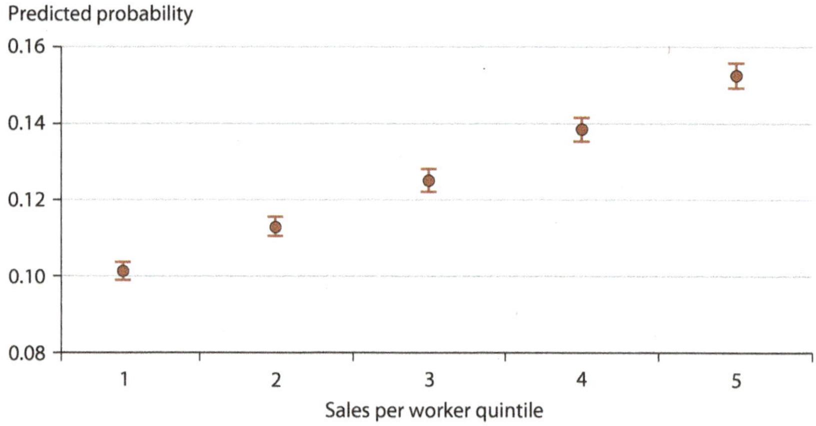

Source: Calculations based on the World Bank Firms and Climate Adaptation Policy Toolkit (World Bank 2025) using World Bank Enterprise Surveys (WBES) data.
Note: The WBES sample includes almost 6,000 firms across Egypt, Arab Rep.; Jordan; Lebanon; Morocco; and Tunisia between 2019 and 2020. Results based on probit regressions—dummy for adopting adaptation regressed on sales per worker quintile, country and sector fixed effects, and firm age and size controls. Each circle is the average predicted probability across all firms in the quintile. Error bars show the 95 percent confidence intervals. MENAAP = Middle East, North Africa, Afghanistan, and Pakistan.

Table 1F.3 shows regression results of the relationship between firms' investment and adaptation practices in MENAAP.

TABLE 1F.3

Relationship Between Firms' Investment and Adaptation Practices, MENAAP Countries

Dependent variable	Adoption of heating or cooling technology (dummy)	Monitoring water usage (dummy)	Monitoring emissions (dummy)
Investment in fixed assets (dummy)	0.608*** (0.056)	0.351*** (0.079)	0.335* (0.193)
Obs	4,636	3,071	322
Country and sector FE	Yes	Yes	Yes
Year FE	No	No	No
Country-sector FE	No	No	No
Controls	Yes	Yes	Yes

Source: Table original to this publication. Calculations based on the World Bank Firms and Climate Adaptation Policy Toolkit (World Bank 2025) using World Bank Enterprise Surveys (WBES) data.
Note: The WBES sample includes almost 6,000 firms across Egypt, Arab Rep.; Jordan; Lebanon; Morocco; and Tunisia between 2019 and 2020. Probit regressions predict the likelihood of adopting adaptation practices. Table displays base coefficients. FE = fixed effects; MENAAP = Middle East, North Africa, Afghanistan, and Pakistan.
Significance level: * = 10 percent, *** = 1 percent.

ANNEX 1G. ENERGY EFFICIENCY RESULTS: SUPPLEMENTARY FIGURE AND TABLES

Figure 1G.1 presents the energy intensity of the manufacturing sector in MENAAP countries, while table 1G.1 examines the effect of firm characteristics on firm energy intensity, and table 1G.2 presents the effect of extreme heat days on firm electricity consumption in MENAAP countries.

FIGURE 1G.1

Energy Intensity of the Manufacturing Sector in MENAAP Countries Over Time

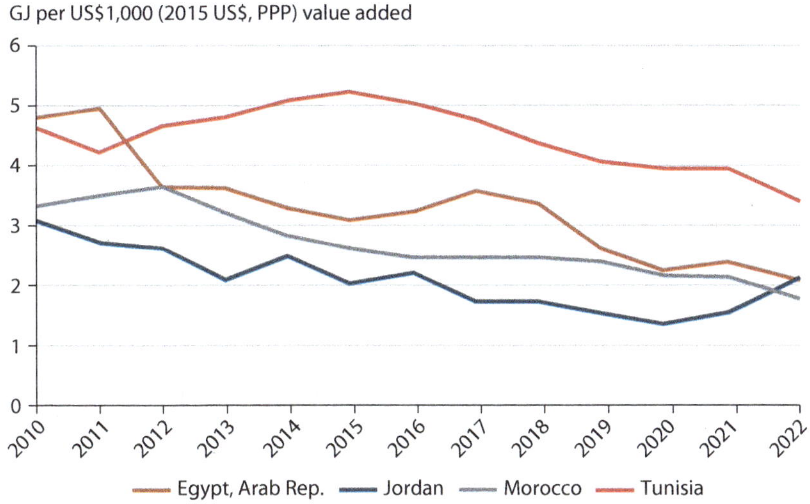

Source: Calculations based on the World Bank Firms and Climate Mitigation Policy Toolkit (World Bank 2025) using International Energy Agency (IEA) and World Development Indicators (WDI) estimates.

Note: Manufacturing energy consumption is based on end users' consumption data from the IEA. This corresponds to International Standard Industrial Classification (ISIC) rev.4 10–18 and 20–32 divisions. Value-added data are taken from the WDI "Manufacturing, value added" series in current local currency units (indicator NV.IND.MANF.CN), which are then adjusted for purchasing power parity (PPP) using WDI conversion factors (indicator PA.NUS.PPP). Manufacturing in WDI data corresponds to ISIC rev.3.1 15–37 divisions. GJ = gigajoules; MENAAP = Middle East, North Africa, Afghanistan, and Pakistan.

TABLE 1G.1

Effect of Firm Characteristics on Firm Energy Intensity, MENAAP Countries

Dependent variable	Electricity cost as a share of sales revenue (log)				
Effect of firm characteristic	Sales per worker (log) −0.618*** (0.011)	GVC participant −0.200*** (0.034)	Small (compared to large) 0.935*** (0.047)	Medium (compared to large) 0.460*** (0.048)	Managerial experience −0.004** (0.002)
Obs	9,011	9,026	9,026	9,026	8,794
R2	0.358	0.123	0.119	0.119	0.119

Source: Table original to this publication. Calculations based on the World Bank Firms and Climate Mitigation Policy Toolkit (World Bank 2025) using World Bank Enterprise Surveys (WBES).
Note: World Bank Enterprise Surveys (WBES) sample includes over 10,000 firms across Egypt, Arab Rep.; Iraq; Jordan; Lebanon; Morocco; Saudi Arabia; and Tunisia between 2019 and 2024. All regressions include country, sector, year, and country-sector fixed effects and firm size and age controls. Regressions are not weighted. Effect on electricity intensity not significant for innovator. Small firms are those with 5–19 employees; medium firms, 20–99 employees; and large firms, 100 or more employees. GVC = global value chain; MENAAP = Middle East, North Africa, Afghanistan, and Pakistan. Significance level: ** = 5 percent, *** = 1 percent.

TABLE 1G.2

Effect of Extreme Heat Days on Firm Electricity Consumption, MENAAP Countries

Dependent variable	Electricity consumption in kWh (log)	Share of electricity (%) produced by own generator (log)
Heat days (dev LT trend)	−0.286*** (0.064)	0.168*** (0.041)
Obs	1,619	1,429
R2	0.346	0.461

Source: Table original to this publication. Calculations based on the World Bank Firms and Climate Adaptation Policy Toolkit (World Bank, 2025) using World Bank Enterprise Surveys (WBES) and Database of Regional Climate Change data (Bettarelli et al., 2024).
Note: "Heat days" are those whose maximum temperature exceeds 35 degrees Celsius. The middle column runs on 6,000 firms across Egypt, Arab Rep.; Jordan; Lebanon; Morocco; and Tunisia covering 2019 to 2020 (surveys from the World Bank Enterprise Surveys Green Module). The right column runs on the full sample of countries, with over 10,000 firms across Egypt, Arab Rep., Iraq, Jordan, Lebanon, Morocco, Saudi Arabia, and Tunisia, covering 2019 to 2024. All regressions include country, sector, year, and country-sector fixed effects and firm size and age controls. Regressions are not weighted. kWh = kilowatt-hours; LT = long-term; MENAAP = Middle East, North Africa, Afghanistan, and Pakistan. Significance level: *** =1 percent.

ANNEX 1H. GREEN TECHNOLOGY ADOPTION IN MOROCCO

Recent data from the World Bank's 2024 Firm-level Adoption of Technologies (FAT) survey of 1,256 Moroccan establishments reveals a mixed picture of green technology adoption (Cirera et al. 2026). Morocco demonstrates some progress in green technology adoption, but with distinctive characteristics.

Key Facts

Morocco shows relatively higher adoption of captive renewable energy (6 percent) than peer countries, surpassing India (3.4 percent) and Viet Nam (0.5 percent). Morocco's adoption of energy-efficient technologies compares favorably with other low- and middle-income economies, with Energy Star certification at 28.7 percent; energy-efficient lighting at 65.1 percent; and variable air volume (VAV) and/or heating, ventilation, and air-conditioning (HVAC) systems at 19.2 percent.

In terms of environmental management, International Organization for Standardization (ISO) 14001 certification has been adopted by approximately 15 percent of firms, but other green management practices remain limited. Only 5 percent of firms monitor carbon dioxide (CO_2) and/or carbon monoxide (CO) emissions, and less than 3 percent participate in carbon trading. Water resource management presents a significant challenge, as more than 93 percent of establishments do not treat wastewater.

Barriers and Drivers

Several factors influence technology adoption in Moroccan firms. Competition emerges as the primary driver (cited by approximately 30 percent of firms), followed by equipment depreciation and replacement needs. Cost constraints represent the leading barrier (cited by 25–30 percent of firms), followed by lack of finance and uncertain demand.

Management quality plays a significant role, with firms led by managers who studied abroad (27.9 percent overall, increasing to 49.2 percent for large firms) showing higher technology sophistication. International exposure also correlates with advancement, because firms engaged in exports or connected to multinational companies demonstrate higher technology adoption rates. Interestingly, gender dynamics appear relevant, with establishments led by female top managers (only 4.2 percent of all firms) showing higher adoption of general business function technologies.

As Morocco continues its digital transformation, targeted policies addressing these barriers and leveraging existing drivers could accelerate both digital and green technology adoption, particularly among smaller firms where the technology gap is most pronounced.

ANNEX 1I. BUSINESS ENVIRONMENT RESULTS: SUPPLEMENTARY FIGURES AND TABLES

TABLE 1I.1

Effects of Extreme Heat Days Interacted with Business Environment on Sales, MENAAP Countries

Dependent variable	Sales	
Heat days (dev LT trend)	−0.060*** (0.023)	−0.028 (0.022)
Interaction	Regulation index (poorer) −0.041*** (0.019)	Access to finance index (poorer) −0.125*** (0.020)
Obs	9,411	9,369
R2	0.666	0.685

Source: Table original to this publication. Calculations based on the World Bank Firms and Climate Adaptation Policy Toolkit (World Bank 2025) using World Bank Enterprise Surveys (WBES) and Database of Regional Climate Change data (Bettarelli et al. 2024).

Note: "Extreme heat" days are those whose maximum temperature exceeds 35 degrees Celsius. Geocoded satellite weather data were matched with geocoded WBES data for over 10,000 firms across Egypt, Arab Rep.; Iraq; Jordan; Lebanon; Morocco; Saudi Arabia; and Tunisia between 2019 and 2024. All regressions include controls for long-term (LT) means of heat days, country, sector, year, and country-sector fixed effects, and firm size and age controls. Regressions are not weighted. The interaction between heat days and the labor force index is not significant. MENAAP = Middle East, North Africa, Afghanistan, and Pakistan.

Significance level: *** = 1 percent.

FIGURE 1I.1

Marginal Effect of Extreme Heat Days on Firm Sales at Varying Levels of Regulatory Constraints, MENAAP Countries

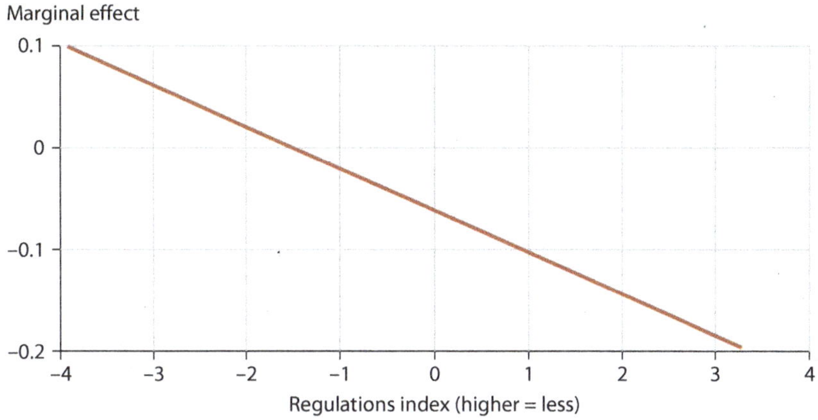

Source: Calculations based on the World Bank Firms and Climate Adaptation Policy Toolkit (World Bank 2025) using World Bank Enterprise Surveys (WBES) and Database of Regional Climate Change (DRCC) data (Bettarelli et al. 2024).
Note: "Extreme heat" days are those whose maximum temperature exceeds 35 degrees Celsius. Geocoded satellite weather data were matched with geocoded WBES data for over 10,000 firms across Egypt, Arab Rep.; Iraq; Jordan; Lebanon; Morocco; Saudi Arabia; and Tunisia between 2019 and 2024. All regressions include controls for long-term means of heat days, country, sector, year, and country-sector fixed effects, and firm size and age controls. Regressions are not weighted. MENAAP = Middle East, North Africa, Afghanistan, and Pakistan.

TABLE 11.2

Effects of Extreme Heat Days Interacted with Business Environment Conditions on Sales per Worker, MENAAP Countries

Dependent variable	Sales per worker	
Heat days (dev LT trend)	−0.035* (0.018)	−0.012 (0.018)
Interaction	Regulation index (poorer) 0.006 (0.016)	Access to finance index (poorer) −0.118*** (0.017)
Obs	9,385	9,343
R2	0.724	0.736

Source: Table original to this publication. Calculations based on the World Bank Firms and Climate Adaptation Policy Toolkit (World Bank 2025) using World Bank Enterprise Surveys (WBES) and Database of Regional Climate Change data (Bettarelli et al. 2024).
Note: "Extreme heat" days are those whose maximum temperature exceeds 35 degrees Celsius. Geocoded satellite weather data were matched with geocoded WBES data for over 10,000 firms across Egypt, Arab Rep.; Iraq; Jordan; Lebanon; Morocco; Saudi Arabia; and Tunisia between 2019 and 2024. All regressions include controls for long-term (LT) means of heat days, country, sector, year, and country-sector fixed effects, and firm size and age controls. Regressions are not weighted. The interaction between heat days and the labor force index is not significant. MENAAP = Middle East, North Africa, Afghanistan, and Pakistan.
Significance level: * = 10 percent, *** = 1 percent.

Appendix B: Annexes for Chapter 2

ANNEX 2A. THE UK CARBON BORDER ADJUSTMENT MECHANISM

Following the example of the European Union (EU), some high-income economies are planning to install a border carbon adjustment or a carbon tax for imported goods. Canada, Japan, and the United Kingdom (UK) are all contemplating the creation of their own carbon border adjustments (Burnett et al. 2026).

The UK Carbon Border Adjustment Mechanism (CBAM) is a proposed policy tool designed to address the risk of carbon leakage and ensure a level playing field for industries subject to the UK's climate policies. It does so by taxing imports of carbon-intensive goods, ensuring these imports face a carbon price comparable to that of domestically produced goods. In December 2023, the UK government announced its intention to implement a CBAM as of January 1, 2027, to specific carbon-intensive goods (DESNZ 2023).

Although the UK CBAM draws heavily from the EU CBAM, it exhibits some slight differences (refer to table 2A.1):

- The UK CBAM covers aluminum, cement, fertilizer, hydrogen, and iron and steel (like the EU CBAM), but it excludes electricity.
- The UK CBAM applies to both direct carbon dioxide (CO_2) emissions from production (scope 1) and indirect CO_2 emissions from energy generation (scope 2) for all sectors (ICAP 2024). In contrast, the EU CBAM applies only to direct emissions for iron and steel and aluminum while covering both direct and indirect emissions for other sectors.
- The carbon price under the UK CBAM varies by sector, whereas the EU CBAM applies a uniform carbon price across all covered sectors.

However, just as in the EU CBAM, if a carbon price is applied in the country of origin for imported CBAM goods, UK importers will only need to pay the difference between the UK CBAM carbon price and the carbon price in the place of origin. Also like the EU CBAM, the scope of the UK CBAM will be expanded in the future. Products from the glass and ceramic sectors that also give rise to a risk of carbon leakage will be considered for future inclusion in the UK CBAM.

TABLE 2A.1

Comparison of EU and UK CBAMs

Provision	EU CBAM	UK CBAM
Sectoral coverage	Aluminum, cement, fertilizer, hydrogen, iron and steel, and electricity sectors	Aluminum, cement, fertilizer, hydrogen, and iron and steel sectors from 2027; inclusion of glass and ceramic sectors at a later date
Scope of emissions covered	Direct and indirect emissions for cement, electricity, and fertilizer sectors (scope 1 and 2); direct emissions for iron and steel and aluminum sectors (scope 1)	Direct and indirect emissions for all covered sectors (scope 1 and 2)
Default values	Default values per product	Default values per product; no information on limits
CBAM rate	CBAM certificate price; universal across all CBAM sectors; reflects EU allowance (EUA) price, to be adjusted weekly	Set separately for each CBAM sector; reflects UK allowance (UKA) price and carbon price support, to be adjusted quarterly
Adjustment for third-country carbon price	Only explicit carbon prices to be recognized	Only explicit carbon prices to be recognized
Minimum threshold for liable importers	50 tons of covered goods or goods with 100 tons of CO_2	£50,000 over a 12-month rolling period
Exemption	Not available	Not available

Source: ICAP 2024.
Note: The European Union Allowance (EUA) price refers to the cost of a carbon credit that allows a company to emit a specified amount of carbon dioxide (CO_2) equivalent within the EU (European Union). The same applies to the UK allowance. CBAM = Carbon Border Adjustment Mechanism.

The CBAM Trade Exposure Index (TEI) indicates that the UK CBAM will pose challenges for specific sectors in some MENAAP countries. Figure 2A.1 presents the index values for aluminum (panel c), and fertilizer (panel f), and iron and steel (panel i) across MENAAP countries. These graphs present the CBAM TEI expressed as a percentage for each of the affected sectors, thus capturing excess carbon payments (excess payments compared with the average UK producer) as a share of global exports of the sector. The graphs in panels b, e, and h illustrate the two components of the CBAM TEI: the horizontal axis illustrates export exposure to the UK market while the vertical axis represents excess carbon payments as share of MENAAP exports to the UK.

The CBAM TEI suggests that several MENAAP countries may lose competitiveness in the aluminum, fertilizer, and iron and steel sectors:[1]

- *Saudi Arabia* is highly exposed to the UK CBAM in the aluminum sector. Of the country's aluminum, 4.4 percent heads to the UK, and its excess carbon payment accounts for 0.5 percent of its aluminum exports to the UK (refer to figure 2A.1, panel b).
- *The United Arab Emirates* is highly exposed to the UK CBAM's impacts in the fertilizer sector. Of its fertilizer exports, 0.2 percent is destined to the UK, and its excess carbon payment accounts for 34 percent of its fertilizer exports to the UK (refer to figure 2A.1, panel e).
- *Tunisia* is most exposed in the iron and steel sector, as 5 percent of Tunisia's iron and steel exports go to the UK, with an excess carbon payment that is equivalent to 26 percent of its exports to the UK (refer to figure 2A.1, panel h).

On the other hand, the UK CBAM could *improve* the competitiveness of Oman in the aluminum sector. This positions its exporters to gain a competitive edge in the UK market, provided they can demonstrate their carbon performance. Oman is likely to gain competitiveness in the UK's aluminum market because its CBAM trade exposure index is negative, which means its carbon emission intensity is lower than average producers in the UK (refer to panel 2A.1, panel c).

FIGURE 2A.1

Some MENAAP Countries' Trade Exposure in Certain UK CBAM Sectors Is Significant

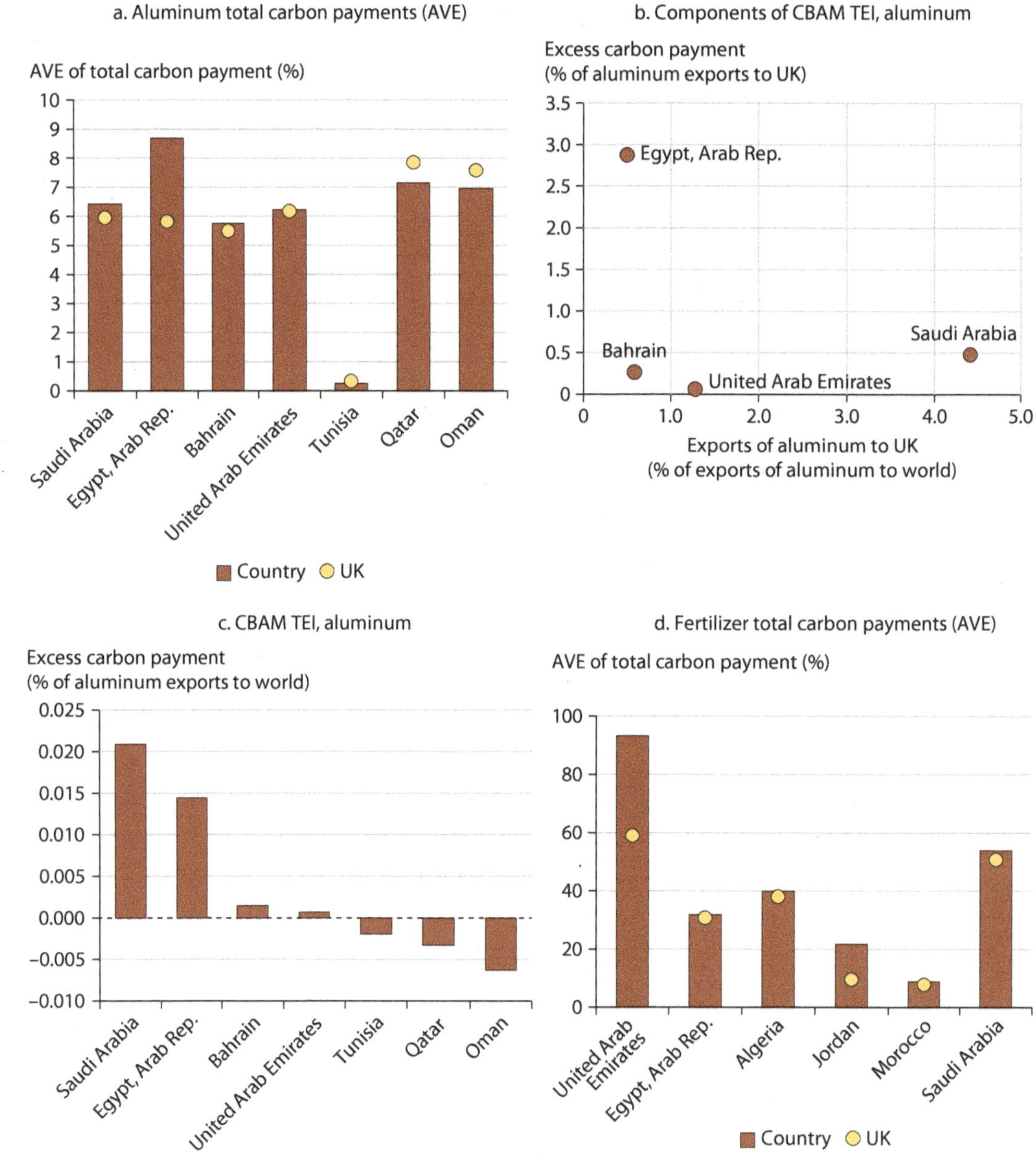

(continued)

FIGURE 2A.1

Some MENAAP Countries' Trade Exposure in Certain UK CBAM Sectors Is Significant (*continued*)

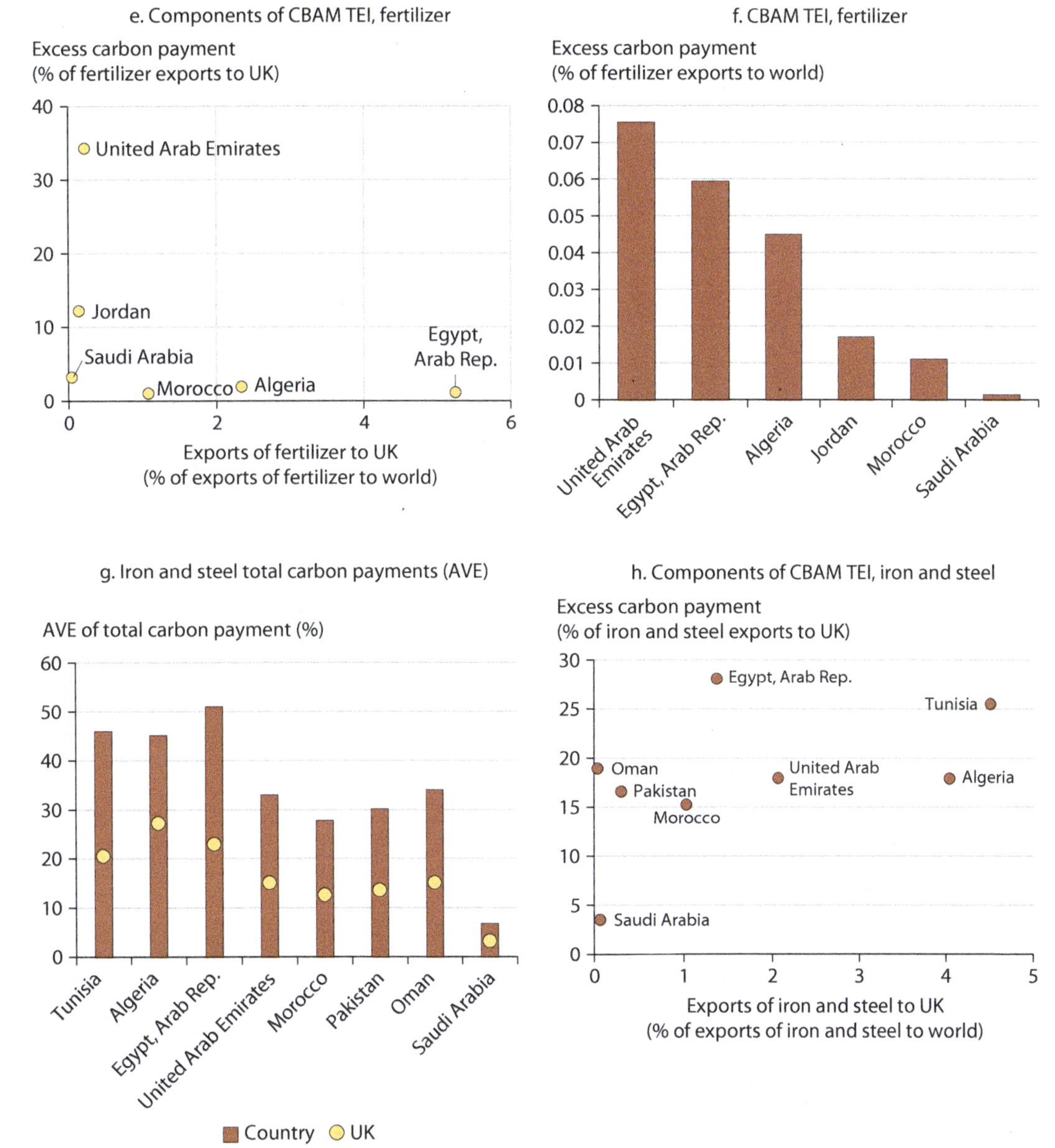

(continued)

FIGURE 2A.1

Some MENAAP Countries' Trade Exposure in Certain UK CBAM Sectors Is Significant (*continued*)

i. CBAM TEI, iron and steel

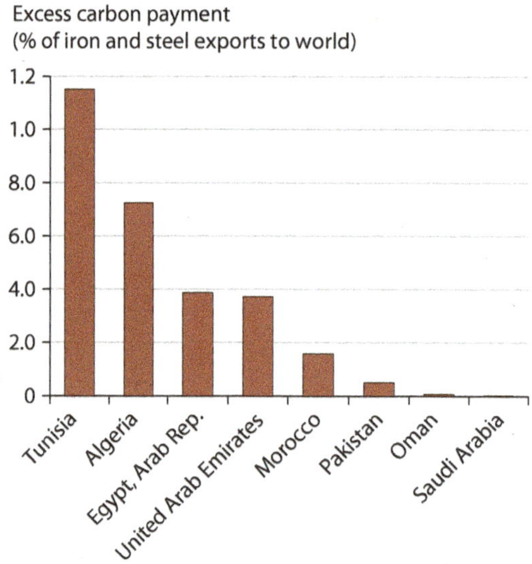

Sources: World Bank calculations using Maliszewska et al. 2025 based on 2022 trade data from the International Trade Analysis Database (BACI) and 2017 emissions intensity data from Vidovic et al. 2023.
Note: Only countries with export values over US$0.5 million are included. Graphs present values for those countries where emission intensity is available. Panels a, d, g, j represent ad valorem equivalent (AVE) of total embedded carbon payments as a share of total exports of these products to the EU. To compare to the embedded carbon payments that an EU producer would have for the same mix of products, the EU average emissions intensity for each product is multiplied by the country's exports by weight to the EU, and aggregated across the relevant products in the sector. Panels b, e, and h present the trade exposure and excess carbon payment in the aluminum, fertilizer, and iron and steel sectors, respectively. They present the decomposition of the Trade Exposure Index (TEI) into two components: carbon intensity relative to the UK and export dependence. Panels c, f, and i express the Carbon Border Adjustment Mechanism (CBAM) TEI in percentage terms, capturing the excess embodied carbon payments as a share of the value of the country's global exports of the affected sector (to achieve this, the TEI value is multiplied by 100). The carbon price is set at US$100 per metric ton of carbon dioxide (CO_2) equivalent, which is the assumed value that the carbon price will reach when the CBAM is fully phased in. Carbon emissions intensity data is converted into kgCO2/US$. AVE = ad-valorem equivalent; CBAM = Carbon Border Adjustment Mechanism; MENAAP = Middle East, North Africa, Afghanistan, and Pakistan.

ANNEX 2B. CARBON PAYMENTS AND EXPORTS OF CBAM-AFFECTED PRODUCTS TO THE UK

TABLE 2B.1

AVE of MENAAP Countries' Total Carbon Payments and CBAM-Affected Exports to the UK

CBAM product	MENAAP country	Exports to UK (% of exports to world)	Exports to UK (US$, millions)	Total carbon payment (% of exports to UK)	UK total carbon payment (% of exports to UK)
Aluminum	Bahrain	0.6	28	5.8	5.5
Aluminum	Egypt, Arab Rep.	0.5	4	8.7	5.8
Aluminum	Oman	1.0	12	7.0	7.6
Aluminum	Qatar	0.5	9	7.1	7.9
Aluminum	Saudi Arabia	4.4	91	6.4	6.0
Aluminum	Tunisia	2.6	4	0.3	0.3
Aluminum	United Arab Emirates	1.3	102	6.2	6.2
Cement	Algeria	0.9	5	165.2	160.3
Fertilizer	Algeria	2.3	42	39.8	37.9
Fertilizer	Egypt, Arab Rep.	5.2	173	31.8	30.7
Fertilizer	Jordan	0.1	1	21.7	9.6
Fertilizer	Morocco	1.1	62	8.8	7.8
Fertilizer	Saudi Arabia	0.0	3	53.9	50.7
Fertilizer	United Arab Emirates	0.2	2	93.1	58.9
Iron and steel	Algeria	4.0	47	45.2	27.3
Iron and steel	Egypt, Arab Rep.	1.4	40	51.0	22.9
Iron and steel	Morocco	1.0	3	27.8	12.6
Iron and steel	Oman	0.0	3	34.0	15.1
Iron and steel	Pakistan	0.3	2	30.1	13.6
Iron and steel	Saudi Arabia	0.1	2	6.7	3.2
Iron and steel	Tunisia	4.5	26	46.0	20.5
Iron and steel	United Arab Emirates	2.1	137	33.0	15.1

Sources: Table original for this publication. Calculations using emission intensities from European Commission (2025) and 2023 trade data from the International Trade Analysis Database (BACI).

Note: The ad-valorem equivalent (AVE) is shown as a percentage of exports to the UK and can be understood as a tariff equivalent of total carbon payments. CBAM = Carbon Border Adjustment Mechanism; MENAAP = Middle East, North Africa, Afghanistan, and Pakistan.

ANNEX 2C. CARBON PAYMENTS AND EXPORTS OF CBAM-AFFECTED PRODUCTS TO THE EU

TABLE 2C.1

AVE of MENAAP Countries' Total Carbon Payments and CBAM-Affected Exports to the EU

CBAM product	MENAAP country	Exports to EU (% of exports to world)	Exports to EU (US$, millions)	Total carbon payment (% of exports to EU)	EU total carbon payment (% of exports to EU)
Aluminum	Bahrain	25	1,184	7	7
Aluminum	Egypt, Arab Rep.	67	521	9	7
Aluminum	Iran, Islamic Rep.	2	13	7	7
Aluminum	Jordan	8	12	2	4
Aluminum	Lebanon	10	5	7	4
Aluminum	Morocco	42	65	1	4
Aluminum	Oman	24	281	7	7
Aluminum	Pakistan	1	2	3	3
Aluminum	Qatar	6	113	6	6
Aluminum	Saudi Arabia	23	477	6	6
Aluminum	Tunisia	55	75	2	2
Aluminum	United Arab Emirates	25	2,001	7	7
Cement	Algeria	23	131	182	176
Cement	Egypt, Arab Rep.	7	60	144	138
Cement	Morocco	22	27	131	184
Cement	Pakistan	0	1	115	76
Cement	Saudi Arabia	2	8	238	220
Cement	Tunisia	37	66	115	109
Cement	United Arab Emirates	1	5	101	77
Fertilizer	Algeria	38	689	36	35
Fertilizer	Bahrain	3	12	50	25
Fertilizer	Egypt, Arab Rep.	38	1,244	29	27
Fertilizer	Jordan	10	98	17	9
Fertilizer	Lebanon	10	2	49	41
Fertilizer	Libya	31	60	32	29
Fertilizer	Morocco	13	743	11	9
Fertilizer	Oman	5	112	47	25
Fertilizer	Qatar	5	96	36	34
Fertilizer	Saudi Arabia	1	68	31	27
Fertilizer	Tunisia	23	69	10	9

(continued)

TABLE 2C.1

AVE of MENAAP Countries' Total Carbon Payments and CBAM-Affected Exports to the EU (*continued*)

CBAM product	MENAAP country	Exports to EU (% of exports to world)	Exports to EU (US$, millions)	Total carbon payment (% of exports to EU)	EU total carbon payment (% of exports to EU)
Fertilizer	United Arab Emirates	3	34	57	27
Iron and steel	Algeria	23	267	44	27
Iron and steel	Bahrain	0	6	25	11
Iron and steel	Egypt, Arab Rep.	41	1,178	52	23
Iron and steel	Iran, Islamic Rep.	2	36	30	28
Iron and steel	Kuwait	3	4	14	7
Iron and steel	Lebanon	15	11	21	9
Iron and steel	Libya	43	142	45	14
Iron and steel	Morocco	38	95	26	15
Iron and steel	Oman	1	81	52	24
Iron and steel	Pakistan	4	20	34	16
Iron and steel	Qatar	2	11	17	8
Iron and steel	Saudi Arabia	6	163	45	21
Iron and steel	Tunisia	56	324	27	13
Iron and steel	United Arab Emirates	7	488	30	14

Sources: Table original for this publication. Calculations using emission intensities from Vidovic et al. 2023 and 2022 trade data from the International Trade Analysis Database (BACI).

Note: The ad-valorem equivalent (AVE) is shown as a percentage of exports to the EU and can be understood as a tariff equivalent of total carbon payments. CBAM = Carbon Border Adjustment Mechanism; EU = European Union; MENAAP = Middle East, North Africa, Afghanistan, and Pakistan.

ANNEX 2D. TRADE DIVERSION OF NON-CBAM PRODUCTS, BY DESTINATION: SUPPLEMENTARY FIGURES

FIGURE 2D.1

Change in Destinations of MENAAP Countries' Exports Relative to Baseline

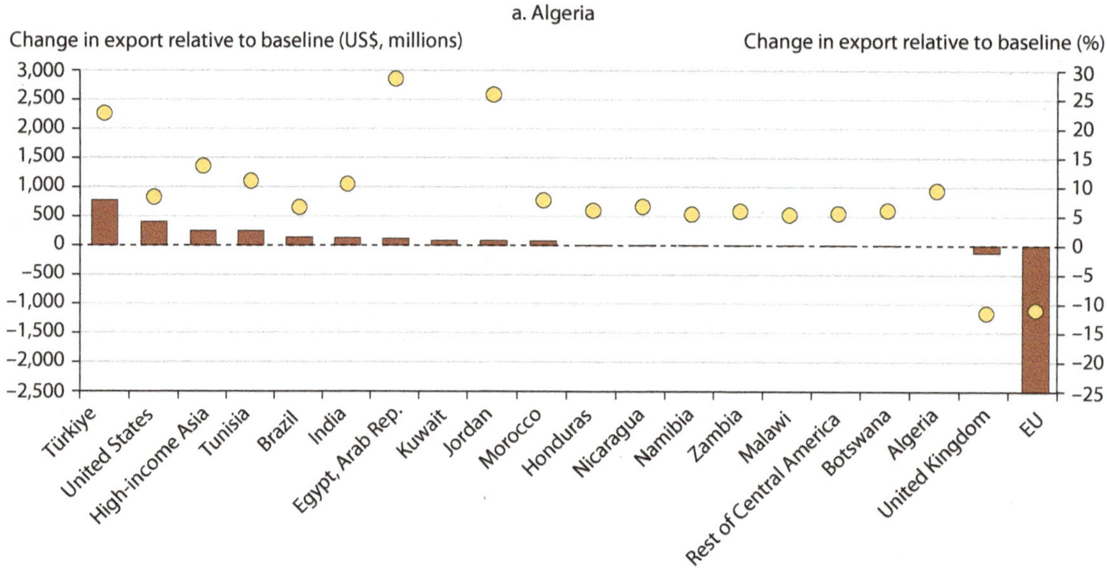

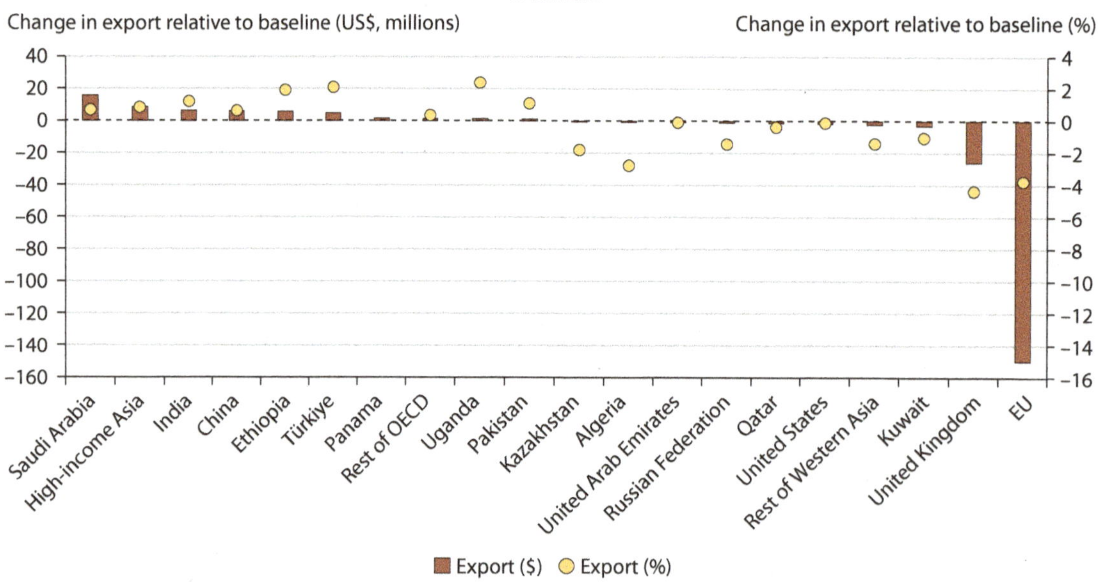

(continued)

FIGURE 2D.1

Change in Destinations of MENAAP Countries' Exports Relative to Baseline (*continued*)

c. Egypt, Arab Rep.

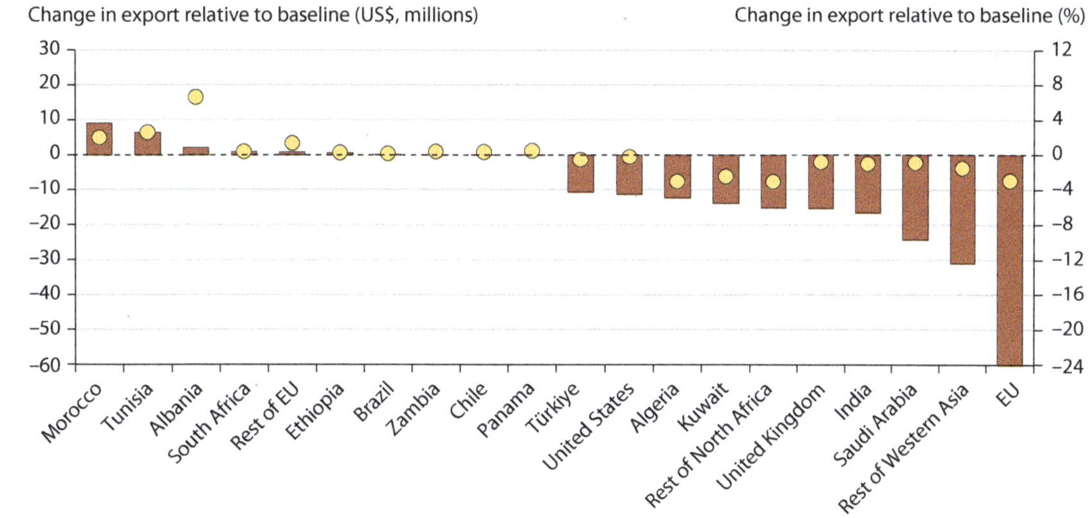

d. Iran, Islamic Rep.

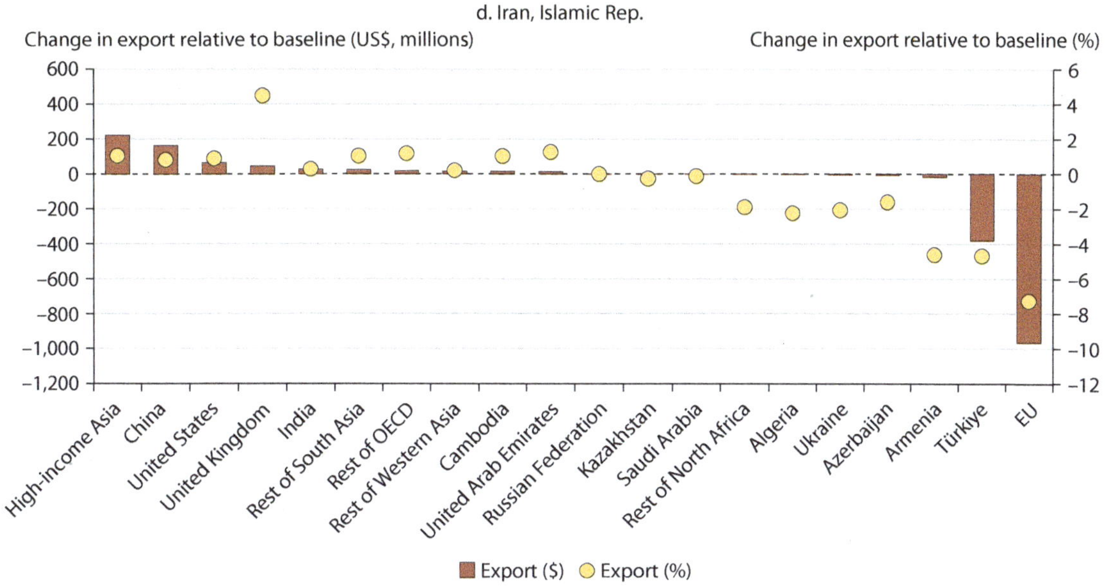

(continued)

FIGURE 2D.1

Change in Destinations of MENAAP Countries' Exports Relative to Baseline (*continued*)

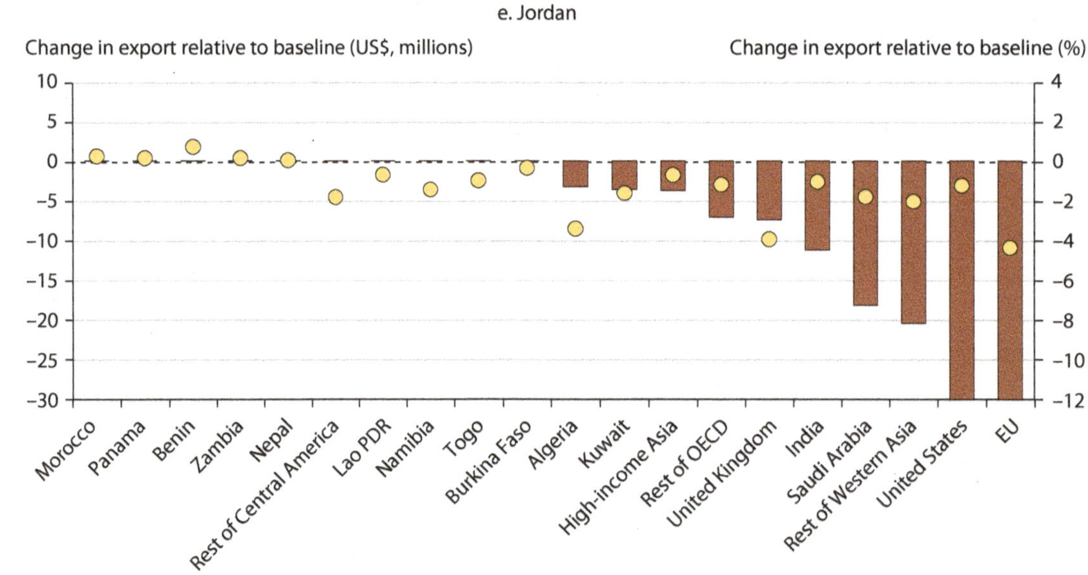

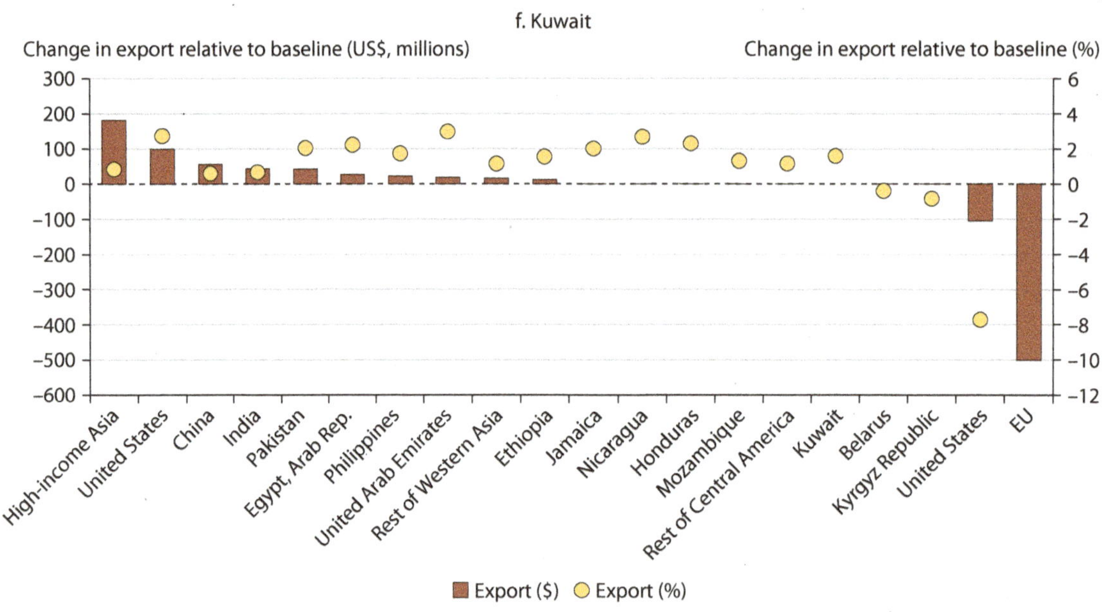

(continued)

FIGURE 2D.1

Change in Destinations of MENAAP Countries' Exports Relative to Baseline (*continued*)

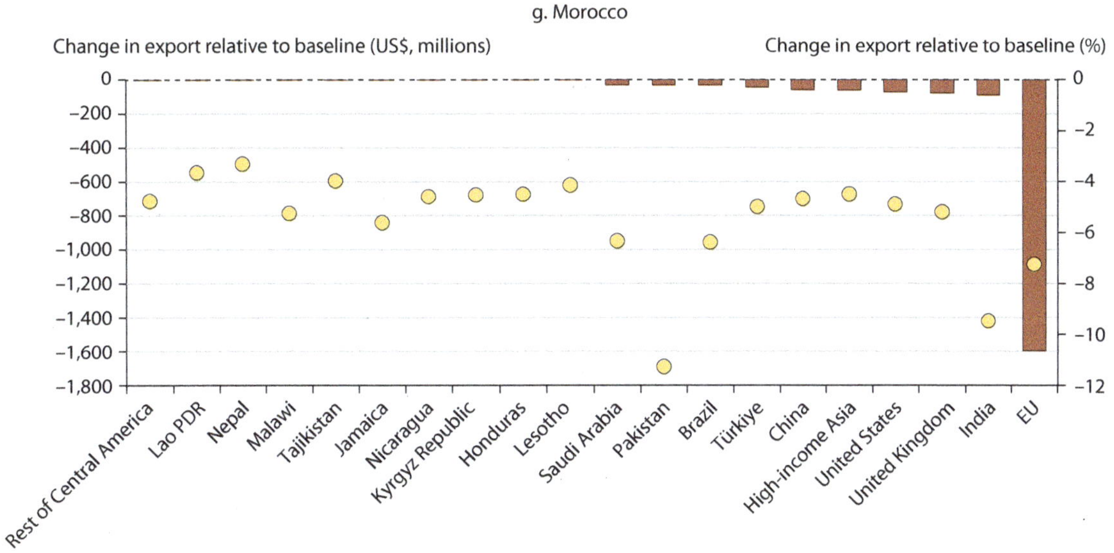

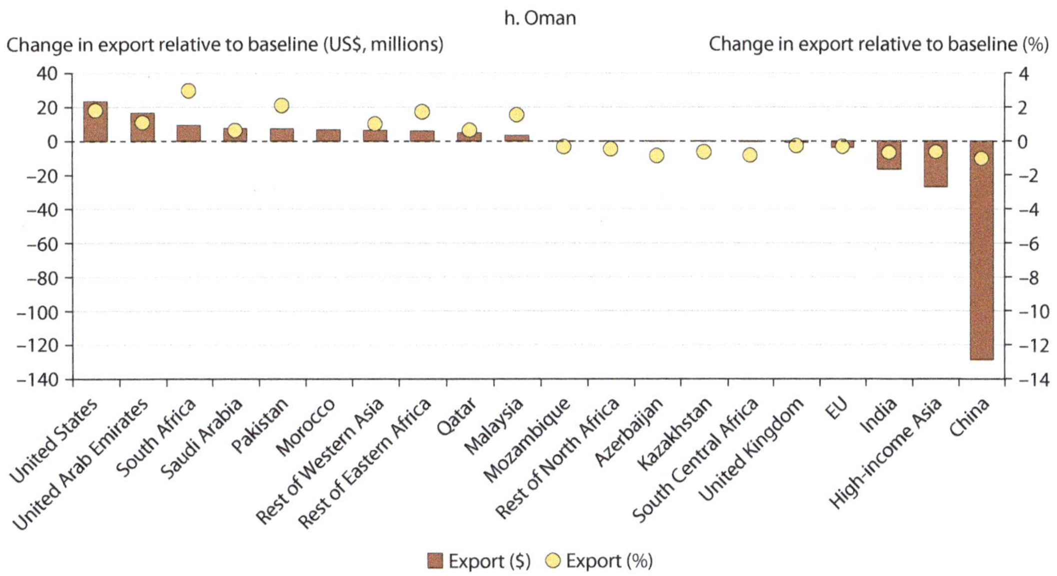

(continued)

FIGURE 2D.1

Change in Destinations of MENAAP Countries' Exports Relative to Baseline (*continued*)

i. Qatar

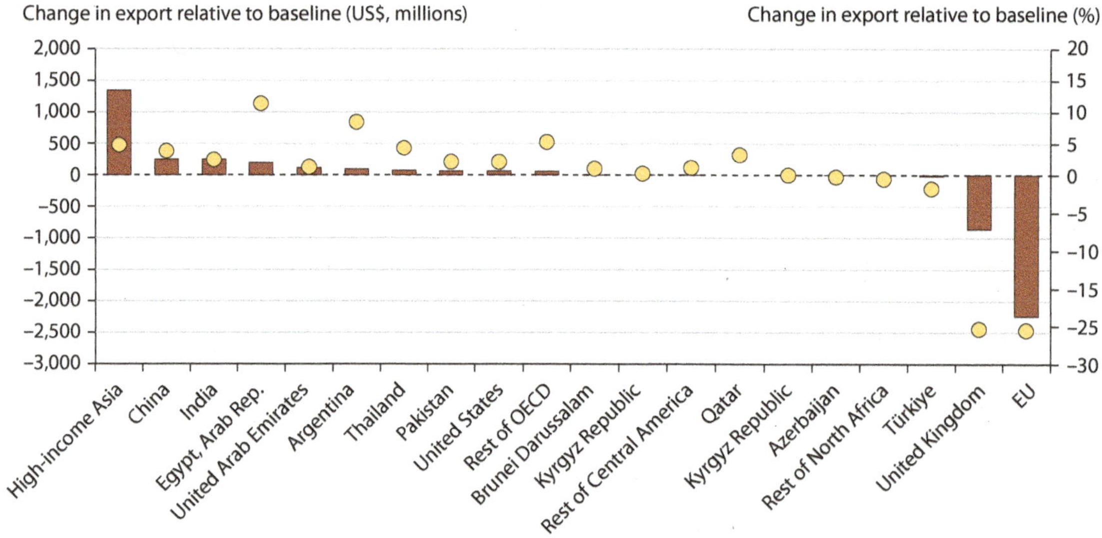

j. Saudi Arabia

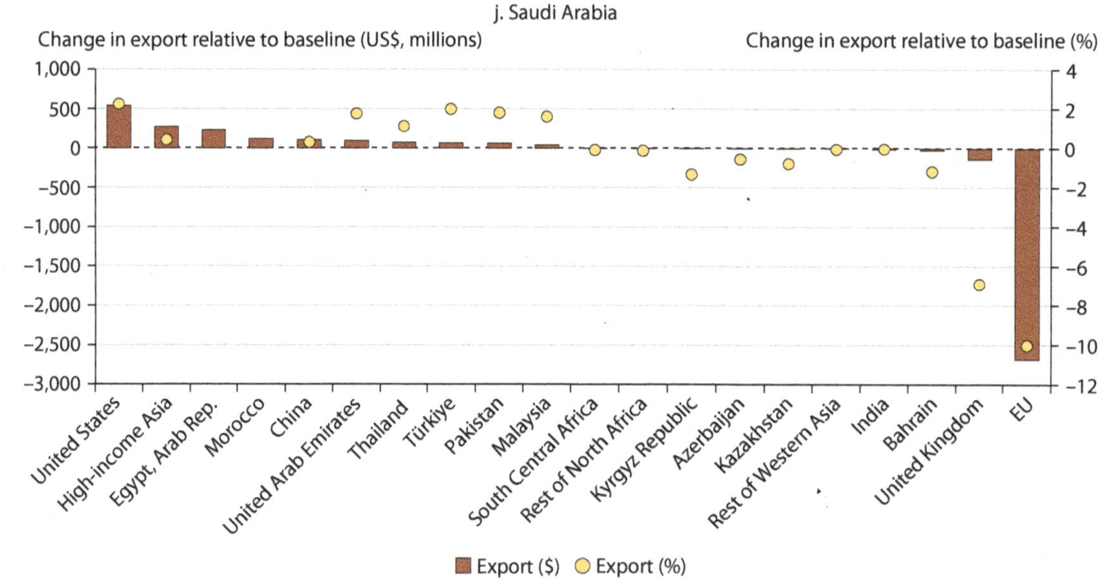

(continued)

FIGURE 2D.1

Change in Destinations of MENAAP Countries' Exports Relative to Baseline (*continued*)

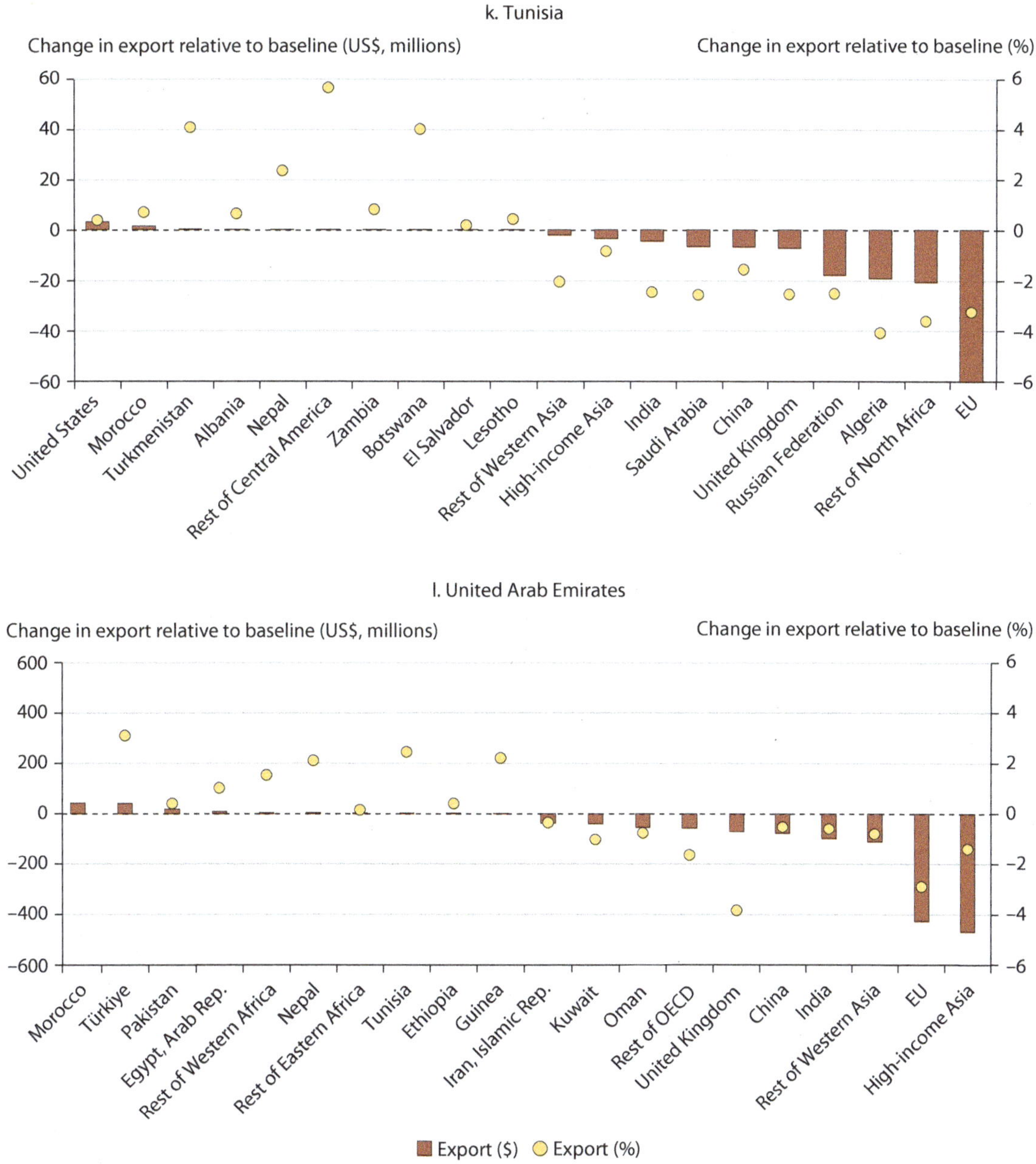

Source: World Bank calculations using Global Trade Analysis Project (Energy) model and GTAP 11 Circular Economy Data Base.

Note: Graphs show the results of computable general equilibrium (CGE) modeling of impacts after the European Union's full Carbon Border Adjustment Mechanism (CBAM) implementation relative to the baseline (business as usual). MENAAP = Middle East, North Africa, Afghanistan, and Pakistan; OECD = Organisation for Economic Co-operation and Development; Rest of Europe = Andorra, Bosnia and Herzegovina, Faroe Islands, Gibraltar, Guernsey, Isle of Man, Jersey, Kosovo, Monaco, Montenegro, North Macedonia, San Marino, Sark, and Vatican City; South Central Africa = Angola and São Tomé and Príncipe.

ANNEX 2E. IMPACT OF EU CBAM ON MENAAP COUNTRIES' EXPORTS

FIGURE 2E.1

Projected Change in MENAAP's Exports to EU and Rest of the World, by Sector

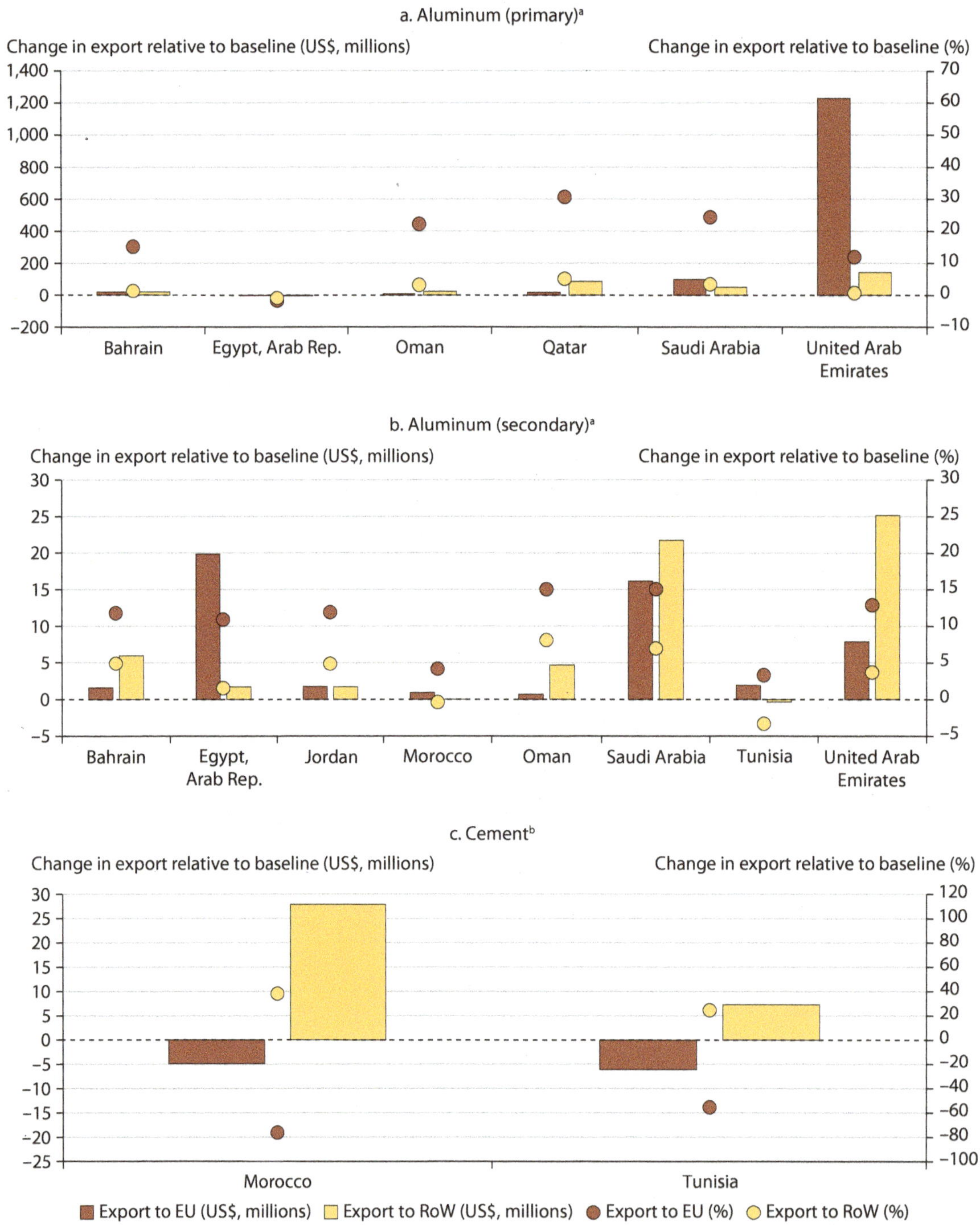

(continued)

FIGURE 2E.1

Projected Change in MENAAP's Exports to EU and Rest of the World, by Sector (*continued*)

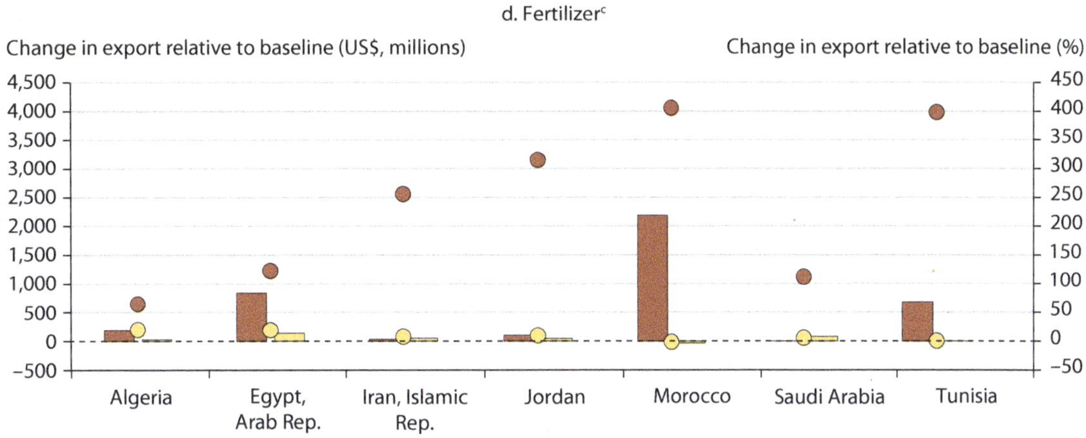

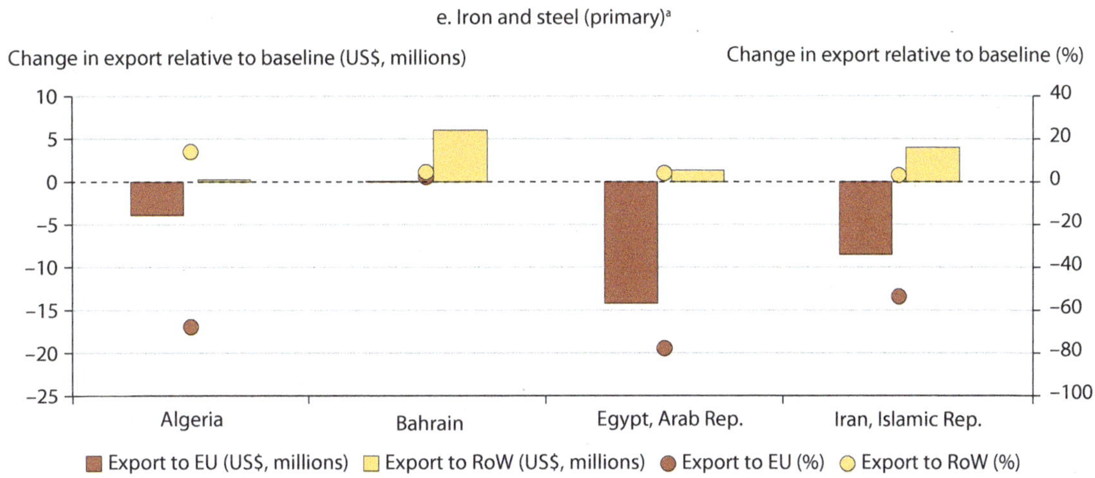

(continued)

FIGURE 2E.1

Projected Change in MENAAP's Exports to EU and Rest of the World, by Sector (*continued*)

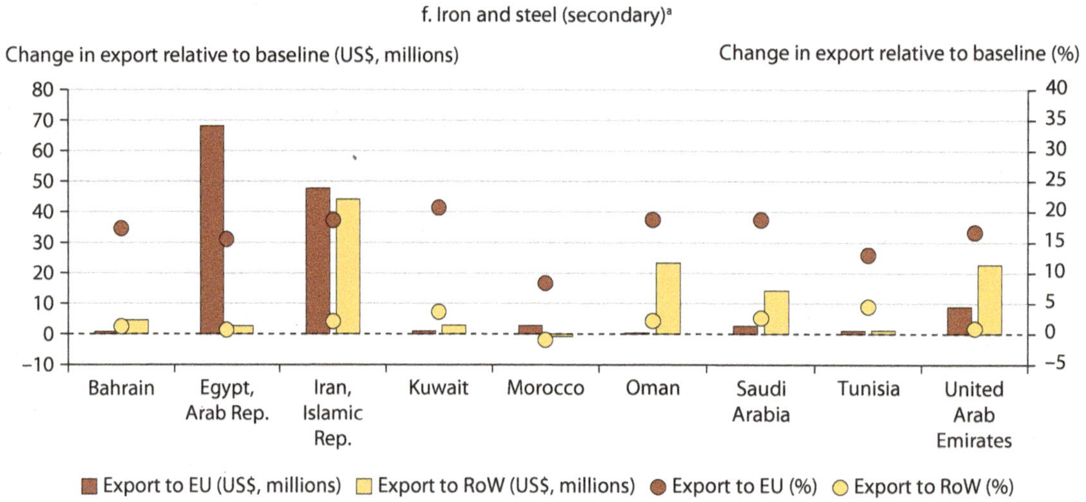

Source: World Bank calculations using Global Trade Analysis Project (Energy) model and GTAP 11 Circular Economy Data Base.

Note: The results are based on the results of computable general equilibrium (CGE) modeling of the impacts of full Carbon Border Adjustment Mechanism (CBAM) implementation. The baseline is business as usual. The bars show changes in US$, millions and the circles, percentage changes. CBAM = Carbon Border Adjustment Mechanism; EU = European Union; MENAAP = Middle East, North Africa, Afghanistan, and Pakistan; RoW = rest of the world.

a. "Aluminum (primary)" and "Iron and steel (primary)" refer to new metal extracted directly from raw natural resources. "Aluminum (secondary)" and "Iron and steel (secondary)" refer to recycled metal produced by melting down existing scrap.

b. We observe more countries increasing their cement exports to the EU under the CBAM because, according to EC (2025) data, high emissions intensities per ton of cement are concentrated in a relatively small group of countries. These high-emitting countries—such as Belarus, Morocco, Türkiye, Ukraine, and a few others—face steeper carbon costs, which significantly reduce their cement exports to the EU. Even if a large number of countries see increases of cement exports into the EU, the overall volume of cement imports into the EU declines, but this decline is concentrated in that small group. In other words, most other exporting countries have relatively lower emissions intensities and, thus, their cement is exported to the EU at prices that are below the average import price into the EU. This gives them a relative advantage under CBAM, allowing them to expand their exports. So, even though total EU imports fall, a wide group of relatively lower-emission, lower-price countries increase their market share.

c. The emission intensities of Algeria and Egypt, Arab Rep., in fertilizer are close to the EU average, making their fertilizer highly competitive. Additionally, factors like fossil fuel price fluctuations, relative emission intensity, and the competitiveness of other sectors affect overall export performance: as some exports decline, others must increase because of a fixed trade balance-to-GDP ratio in simulations.

ANNEX 2F. FDI IN MENAAP

Foreign direct investment (FDI) plays a significant but uneven role in MENAAP's economies. Inward FDI stocks equal approximately half of the region's gross domestic product (GDP) (refer to figure 2F.1), ranking just below Europe and Central Asia among low- and middle-income economies, with considerable differences across countries.

In Lebanon, FDI stocks exceed 180 percent of GDP—largely because of the war and civil strife's impact on output—while Bahrain and Jordan also show high ratios at 96 percent and 80 percent, respectively. In contrast, small oil-rich economies like Kuwait and Qatar have among the lowest FDI-to-GDP ratios, at 10.5 percent and 11.5 percent, respectively.

FIGURE 2F.1

FDI Is Critical for MENAAP's Economies

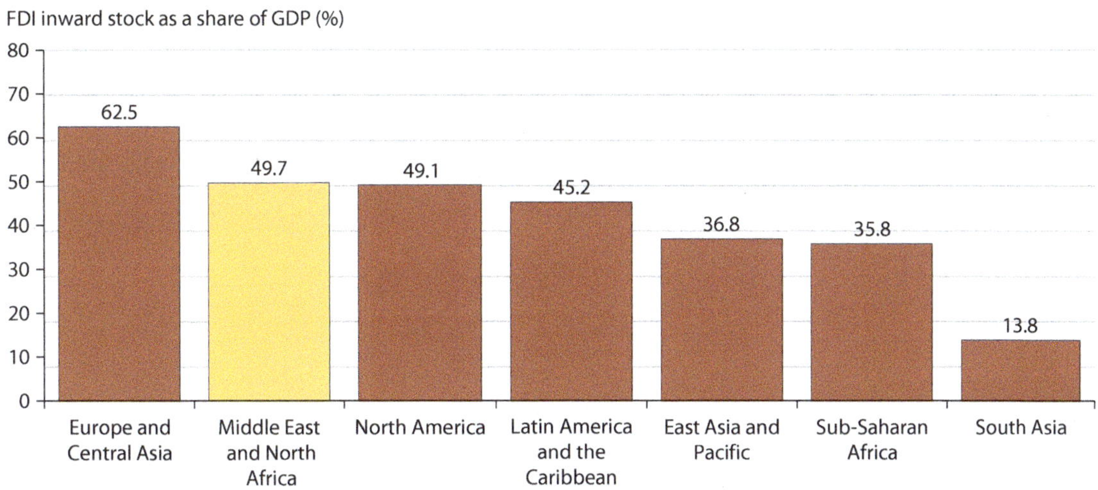

Source: World Bank calculations based on 2023 United Nations Conference on Trade and Development (UNCTAD) Balance of Payment data.
Note: Regional values are calculated by aggregating foreign direct investment (FDI) inward stock (the total accumulated value of all foreign-owned assets at a specific time) and gross domestic product (GDP) across all countries in the region and then dividing the total regional FDI inward stock by the total regional GDP.

However, the MENAAP region has the third lowest share of greenfield FDI in environmentally sustainable projects in emerging markets and developing economies, with significant variation across countries. Green FDI (defined as environmentally sustainable projects based on the EU's taxonomy) accounts for only 15 percent of total greenfield projects in the region—one of the lowest regional shares globally. Only North America (excluding the United States) and East Asia and the Pacific have lower proportions, at 13 percent and 12 percent, respectively (refer to figure 2F.2).

This aggregate figure conceals significant variation across MENAAP countries. Measured either by number of projects or investment in fixed capital assets, the MENAAP countries with the highest proportion of "green" greenfield FDI are the Republic of Yemen, Tunisia, Jordan, Egypt, Djibouti, and Morocco. Those with the lowest shares are Kuwait, Libya, Qatar, and Iraq. The global decrease in greenfield FDI announcements for iron and steel is not influenced by MENAAP; however, significant reductions are observed in cement (table 2F.1).

FIGURE 2F.2

MENAAP Has the Third Lowest Share of "Green" Greenfield FDI among EMDEs

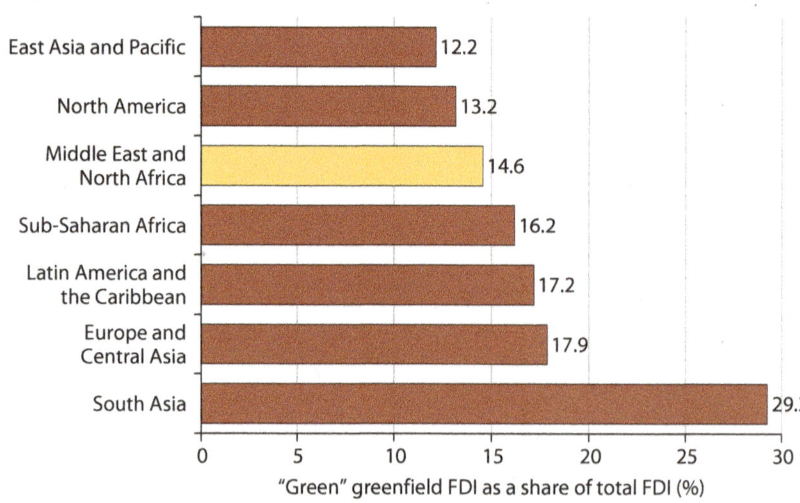

Source: World Bank calculations based on fDi Markets data.
Note: Regional values are calculated by aggregating "green" greenfield foreign direct investment (FDI) projects across all countries in the region and then dividing the total greenfield FDI projects by the total regional GDP, averaged over the latest three-year data period. FDI greenfield projects are categorized as "green" (environmentally sustainable projects) based on the European Union's taxonomy.

TABLE 2F.1

The Global Decrease in Greenfield FDI Announcements for Iron and Steel Is not Influenced by MENAAP, but Significant Reductions Are Observed in Cement

Region	Aluminum	Iron and steel	Fertilizer	Cement
Europe and Central Asia	−0.244 (0.152)	0.077 (0.208)	−0.171 (0.215)	0.054 (0.126)
Middle East and North Africa	**−0.435**** (0.204)	0.213 (0.211)	−0.716 (0.451)	**−0.327*** (0.197)
Latin America and the Caribbean	**−0.710***** (0.210)	0.116 (0.233)		
East Asia and Pacific	−0.285 (0.201)	0.068 (0.374)	−0.030 (0.280)	**−1.025***** (0.381)
Sub-Saharan Africa	**−1.478***** (0.320)	0.371 (0.347)	−0.481 (0.540)	
South Asia	**−1.038***** (0.324)	−0.359 (0.268)		**−4.282***** (0.286)
North America	−0.333 (0.275)	−0.037 (0.221)	**−0.503**** (0.200)	

Source: World Bank calculations based on fDi Markets data.
Note: The table shows the results from a gravity model regression (Yotov et al. 2016) using the Trade Exposure Index (TEI) interacted with post-2019 dummy on log capital expenditure (in US$, millions) by country pair and year, with year, source, destination fixed effects, and region-specific interactions (see chapter 2, box 2.1 for details). A 0.25 estimate implies e^(0.25) − 1 = 0.28% higher greenfield foreign direct investment (FDI) in 2019 relative to pre-2019 level. MENAAP = Middle East, North Africa, Afghanistan, and Pakistan.
Significance level: * = 10 percent, ** = 5 percent, *** = 1 percent.

ANNEX 2G. MENAAP COUNTRIES' GVC EXPOSURE TO THE EU CS3D

FIGURE 2G.1

GVC Exposure in the MENAAP Region to EU CS3D, by Sector

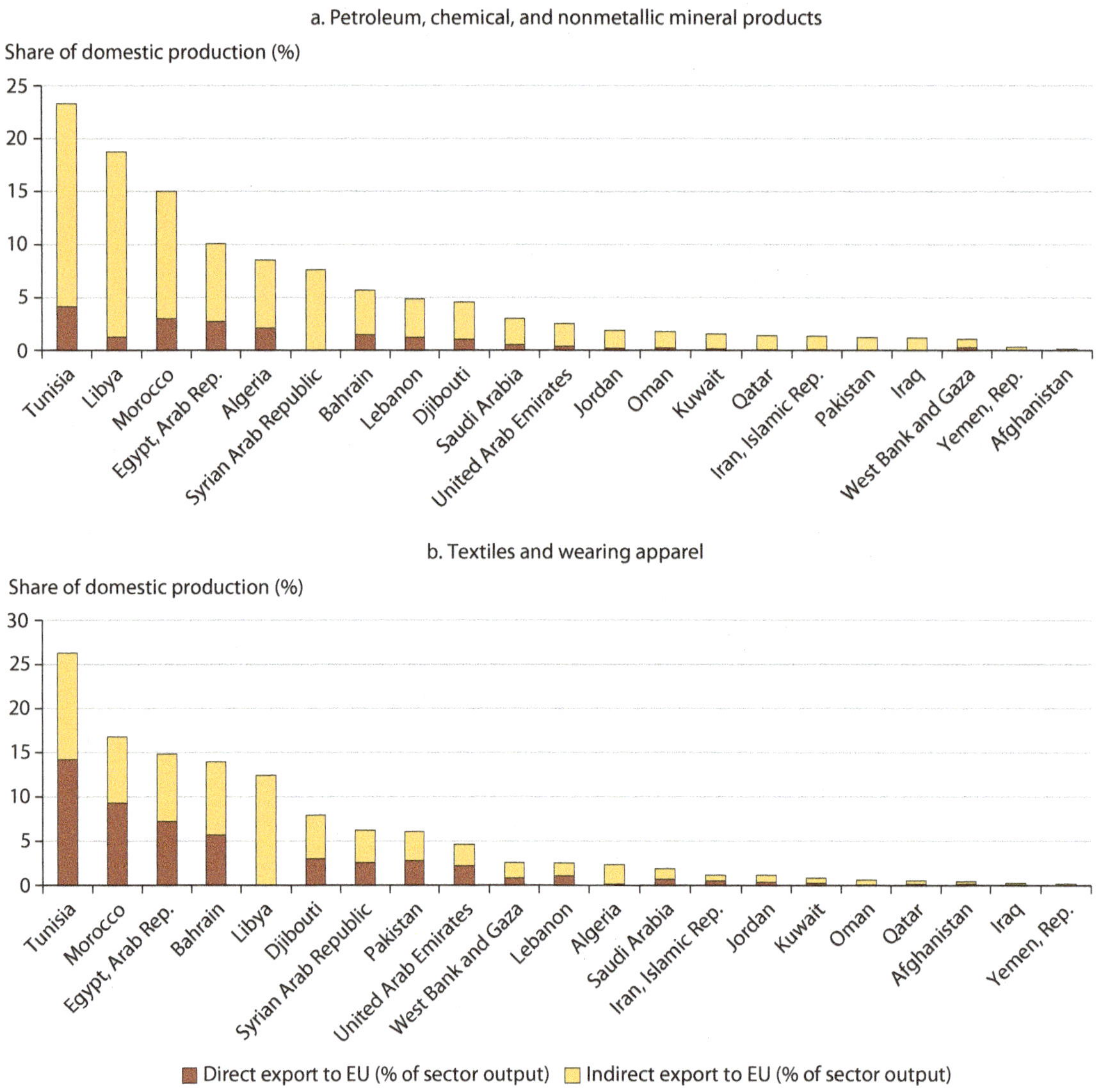

(continued)

FIGURE 2G.1

GVC Exposure in the MENAAP Region to EU CS3D, by Sector (*continued*)

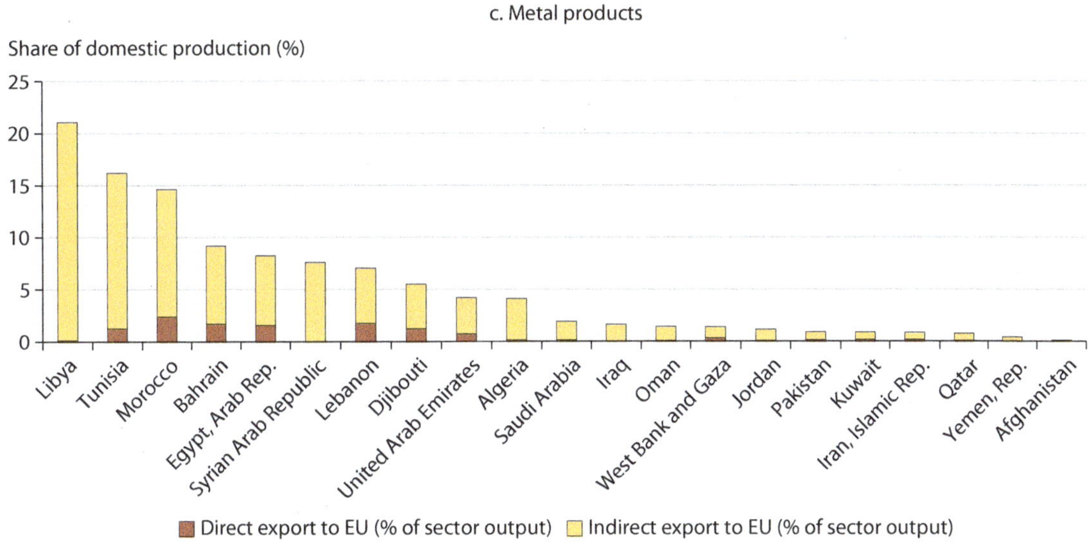

Sources: World Bank calculations and Arvis et al. 2025.
Note: The global value chain (GVC) exposure index follows the indicator developed by Arvis et al. (2025) using the methodology of Baldwin, Freeman, and Theodorakopoulos (2022). Using the 2022 EORA global supply chain database, GVC exposure to the European Union (EU) is the sum of direct exports to the EU and indirect exports to the EU over domestic output of products. CS3D = Corporate Sustainability Due Diligence Directive; MENAAP = Middle East, North Africa, Afghanistan, and Pakistan.

NOTE

1. Algeria faces CBAM trade exposure in cement, with a TEI of 0.04 percent. Of its cement, 0.86 percent is exported to the UK, and excess carbon payments amount to 5 percent of those exports.

REFERENCES

Arvis. J-F., A. Burman, A. Espita, J-C. Mauer, D. Rocha, and D. Ulybina. 2025. "Measuring Exposure and Vulnerability to International Trade Shocks." Policy Research Working Paper 11243, World Bank.

ICAP (International Carbon Action Partnership). 2024. "UK Outlines Details for Carbon Border Adjustment Mechanism Introduction from 2027." ETS News, November 13, 2024. https://icapcarbonaction.com/en/news/uk-outlines-details-carbon-border-adjustment-mechanism-introduction-2027.

Maliszewska, M., C. Fischer, E. Jung, and M. Chepeliev. 2025. "Carbon Border Adjustment Mechanism (CBAM) Exposure Indices Methodological Note." Prosperity Notes Series, Working Paper. 202614, World Bank.

Vidovic, D., L. Zore, J. A. Moya, and A. Marmier. 2023. *Greenhouse Gas Emission Intensities of the Steel, Fertilizers, Aluminum and Cement Industries in the EU and Its Main Trading Partners*. Joint Research Centre Technical Report, Publications Office of the European Union.

Appendix C: Annexes for Chapter 3

ANNEX 3A. SUPPLEMENTARY DATA

FIGURE 3A.1

Share of Firms Citing A2F as a Significant Constraint in MENAAP and Globally

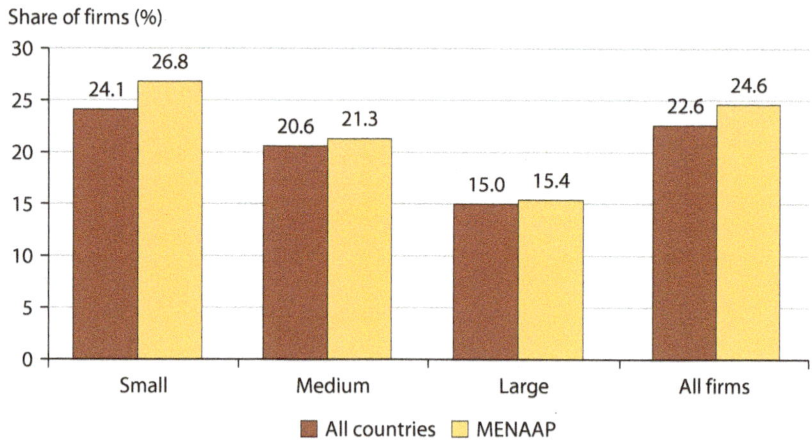

Source: World Bank Enterprise Surveys (WBES) dataset.
Note: Figure shows the share of firms, by size, that reported that access to finance (A2F) is either a "major" or "very severe" business constraint. The WBES data cover 178 countries. Small firms are those with 5–19 employees; medium firms, 20–99 employees; and large firms, 100 or more employees. MENAAP = Middle East, North Africa, Afghanistan, and Pakistan.

FIGURE 3A.2

SMEs' Collateral-to-Loan Value Ratios in MENAAP Countries

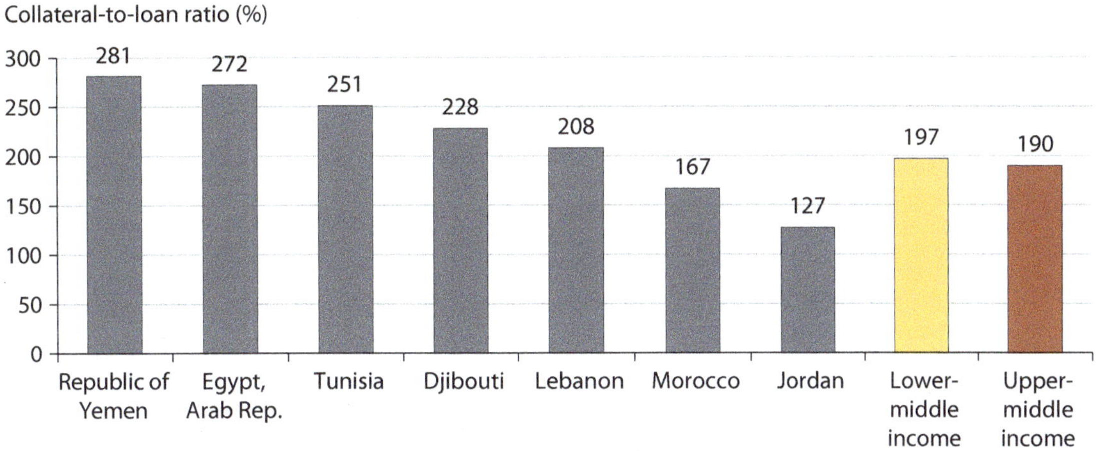

Sources: Betz and Ravasan 2016; Dornel et al. 2020; World Bank estimates.
Note: MENAAP = Middle East, North Africa, Afghanistan, and Pakistan.

FIGURE 3A.3

Issuance of Green Bonds, by MENAAP Country, in Relation to Selected COPs, 2016–25

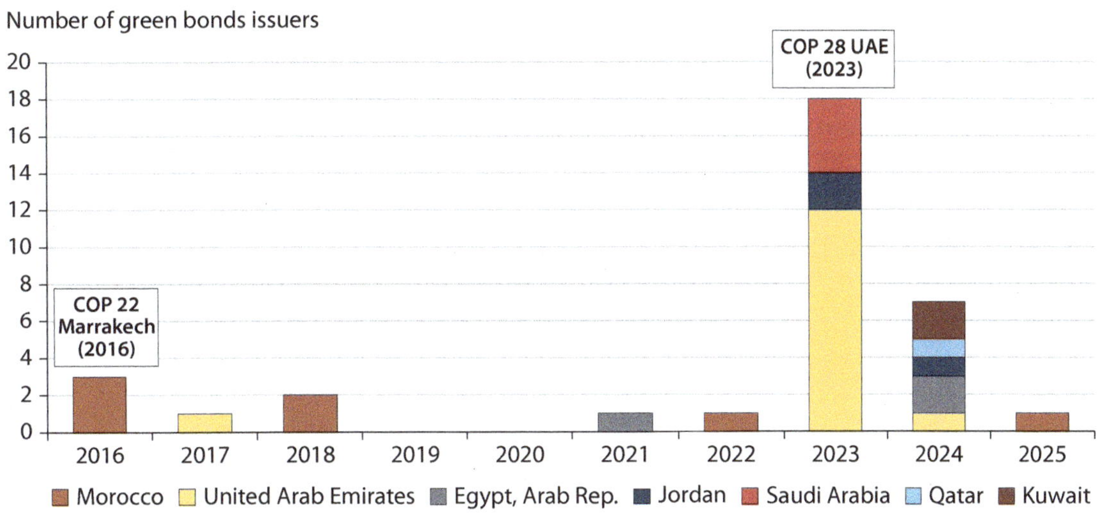

Source: World Bank compilations and data analyses.
Note: The figure displays the number of firms that have issued green bonds over time in the region. COP = Conference of the Parties to the United Nations Framework Convention on Climate Change (UNFCCC); MENAAP = Middle East, North Africa, Afghanistan, and Pakistan; UAE = United Arab Emirates.

FIGURE 3A.4

State Presence in High-Emitting Sectors in Relation to Climate-Vulnerable Sectors in MENAAP Countries

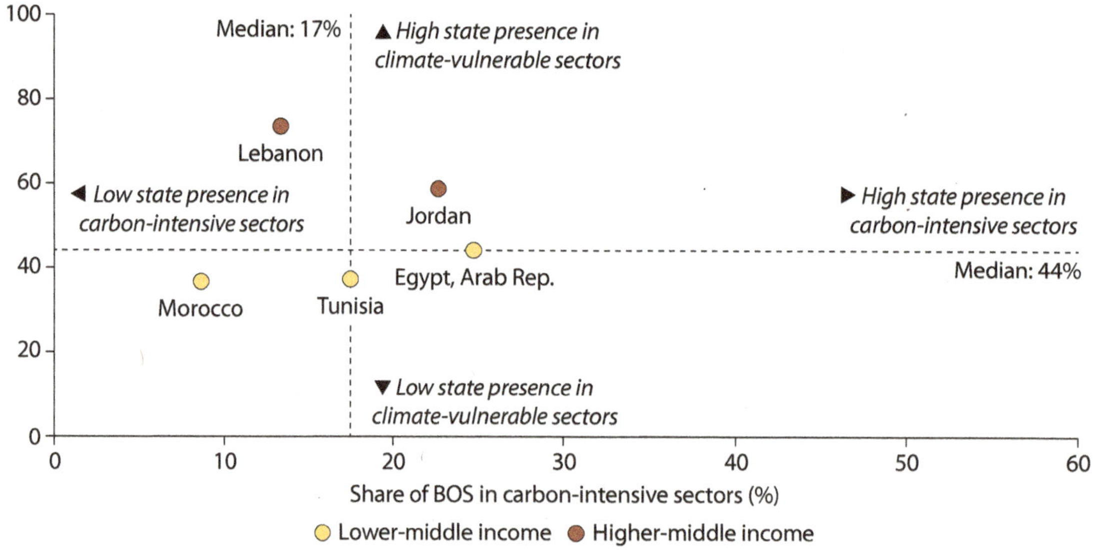

Source: World Bank 2023.

Note: The selected Middle East, North Africa, Afghanistan, and Pakistan (MENAAP) countries are those included in the World Bank's Businesses of the State (BOS) database of firms with direct or indirect state ownership (World Bank 2023). The quadrant thresholds are defined by the respective medians of the selected countries for each axis of the scatterplot. "High-emitting" sectors include mining, oil, gas, and the chemical industry; selected manufacturing activities; selected transportation; selected agricultural activities; and power generation. "Climate-vulnerable" sectors are those most prone to the negative effects of climate change, identified using a risk classification by sectors and converted into Statistical Classification of Economic Activities in the European Community (NACE) code, corroborated with results from the literature. All countries exhibit a large share of BOS in the financial sector and the transportation sector, which are vulnerable to climate. In addition, an important share of BOS operating in "climate-vulnerable sectors" come from the hospitality sector in Jordan, from the construction sector in Morocco and Tunisia, and from the manufacturing and digital sectors in Lebanon. BOS = Businesses of the State.

ANNEX 3B. COUNTRY RESPONSES TO EU CLIMATE MITIGATION POLICIES

As in the Middle East, North Africa, Afghanistan, and Pakistan (MENAAP), many other countries are preparing for upcoming changes in trade regulations. India, Türkiye, and Viet Nam are among those countries.

India's Steps Toward a Domestic Carbon Market

India is taking steps to support its steel and aluminum exports to the EU under CBAM requirements. The government has initiated the foundations for a domestic carbon market to help mitigate CBAM's impact, but businesses, particularly small and medium enterprises (SMEs), continue to face challenges because of limited capacity, complex carbon accounting, and difficulties in accessing funds to innovate. The European Commission plans to help India in adapting to CBAM and developing its emissions pricing system.

Türkiye's Green Deal Action Plan

The Turkish government launched the Green Deal Action Plan (GDAP) in 2021. The aim is to drive a comprehensive green transformation, promoting a circular economy and establishing a national emissions trading system (ETS). A Green Deal Working Group has been set up and is developing methodologies for monitoring emissions, creating a national ETS, and ensuring compliance with EU standards. A pilot ETS targeting high-emission sectors is being prepared.

There is a funding gap for this green transition, with the government seeking external support and developing new incentives for green technologies. For now, Turkish exporters seem to respond to CBAM by shifting focus to less regulated markets, while major companies are investing in carbon reduction technologies.

Viet Nam's Focus on Textile Exports

Viet Nam is working on the EU CS3D. Viet Nam's SMEs face challenges in complying with the CS3D regulations for textile exports. The government developed a National Action Program (2023–27) that aims to promote responsible business practices. Viet Nam's Textile and Apparel Association (VITAS) has teamed up with international organizations to offer training on environmental, social, and governance (ESG) risk management.

Despite these initiatives, 80 percent of the country's SMEs still struggle with meeting CS3D requirements. They face increased costs for compliance, process improvements, improved oversight of subsuppliers, and investments in new management and reporting systems as well as eco-friendly technologies.

ANNEX 3C. ESTIMATING REVENUES WITH CPAT

The Climate Policy Assessment Tool (CPAT) is a tool for analyzing the impacts of carbon pricing and fossil fuel reforms, among other climate policies, along with several economic and noneconomic dimensions (Black et al. 2023). CPAT's main strength is its ability to produce rapid results and its flexibility in policy design. In particular, it enables a comprehensive assessment of the combined impact of a carbon tax and the phasing out of fossil fuel subsidies. Table 3C.1 summarizes the selected main parameters.

CPAT can also recycle the revenues through different options (labor tax cuts, corporate tax cuts, public investment, current spending, and compensatory transfers to households), but this alternative is not considered here.

Revenues from carbon pricing policies are calculated net of indirect changes in revenues (or expenditure) from preexisting energy taxes (or subsidies). Direct revenues from carbon pricing simply correspond to the price of carbon multiplied by the carbon dioxide (CO_2) emissions to which it applies.

TABLE 3C.1

Main Parameters for CPAT Analysis

Parameter	Unit	Comments
Carbon tax scenario		
Carbon tax	US$/$tCO_2$	The carbon tax scenario sets carbon prices at levels required to offset CBAM obligations without accounting for the subsidies.
Policy coverage	Boolean	All fuels, sectors, and industries covered by CPAT are selected.
Total carbon price scenario		
Carbon tax	US$/$tCO_2$	The same tax level as in the previous scenario is used.
Policy coverage	Boolean	All fuels, sectors, and industries covered by CPAT are selected.
Phasing out fossil fuel subsidies, including • Exemptions • Fossil fuel grants • Price controls • VAT rates deviating from the economywide rate applied to fuel	Phasing-out years	All existing types of subsidies are phased out over a period of five years, beginning in 2026.

Source: Black et al. 2023.
Note: CBAM = Carbon Border Adjustment Mechanism; CPAT = Climate Policy Assessment Tool; tCO_2 = tons of carbon dioxide equivalent; VAT = value added tax.

Figure 3C.1 breaks down the various revenue components considered in CPAT. The tool first estimates revenues under the baseline (business-as-usual) scenario. It then calculates revenues under the policy scenario, where tax rates and fuel consumption are adjusted according to the simulated policies. This leads to changes in revenues from existing excise and other taxes, new excise or carbon taxes, value added tax (VAT), and the phaseout of subsidies. Any remaining fossil fuel subsidies in the policy scenario are subtracted from the total estimated revenue.

Additionally, pursuing fossil fuel subsidy reforms can encourage efficiency and discourage overconsumption. Large fossil fuel subsidies in MENAAP undermine the carbon price signal by artificially lowering prices, thereby driving up carbon-intensive output and making exports more emissions intensive than those from the EU.

FIGURE 3C.1

CPAT Revenue and Cost Components

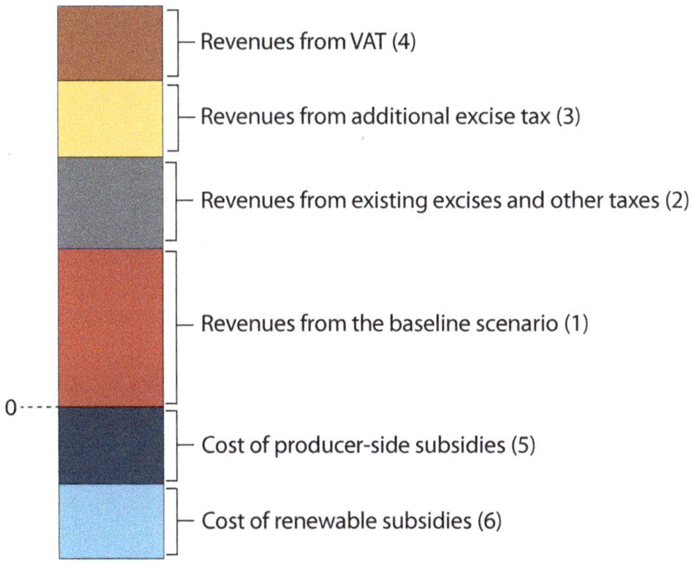

Source: World Bank compilations.
Note: The conceptual chart shows the four revenue components considered within the Climate Policy Assessment Tool (CPAT), which are shown above "0," and the two costs considered in the structure of the tool. The "baseline scenario" represents business as usual. VAT = value added tax.

REFERENCES

Betz, F. and F. R. Ravasan. 2016. "Collateral Regimes and Missing Job Creation in the MENA Region." European investment Bank Working Papers, 2016/03.

Black, S., I. Parry, V. Mylonas, N. Vernon, and K. Zhunussova. 2023. "The IMF–World Bank Climate Policy Assessment Tool (CPAT): A Model to Help Countries Mitigate Climate Change." Working Paper 2023/128, International Monetary Fund.

Dornel, A. D., M. Ait Ali Slimane, and K. Mohindra. 2020. "Improving SME Access to Trade Credit and Financing in MENA." MENA Knowledge and Learning Quick Note 180, World Bank.

World Bank. 2023. "Egypt—Country Climate and Development Report." Report 177494, World Bank.